PREPARE TO DIVE!
The Story of Man Undersea

PREPARE TO DIVE!

The Story of Man Undersea

Written and illustrated by
JACK COGGINS

DODD, MEAD & COMPANY, *New York*

Copyright © 1971 by Jack Coggins
All rights reserved
No part of this book may be reproduced in any form
without permission in writing from the publisher
ISBN 0-396-06384-5
Library of Congress Catalog Card Number: 77-153893
Printed in the United States of America

CONTENTS

THE PIONEERS   7
Early divers—First diving bells—Van Drebbel—De Son—Displacement and buoyancy

BARRELS AND BELLS   15
Lethbridge's cask—Freminet's helmet suit—Symons and Day—Bushnell's *Turtle*—Fulton's *Nautilus*

THEY WALKED UNDER THE WATER   25
Siebe's helmet and helmet suit—Rouquayrol and Denayrouze—Fleuss and the first independent diving apparatus—Effects of pressure—Caisson disease—Nitrogen and "the bends"—Decompression

THE DEADLY FISH   33
Bauer and his submarines—The Confederate "Davids"—Jules Verne—*Le Plongeur*—The torpedo

STEAM AND STEEL BENEATH THE SEA   43
The search for a power plant—Chronology, 1850-1896—Garrett and Nordenfeldt's steam submersibles—Electric motors—Submarines by Ash, Campbell, and Waddington—Goubet's submarines—Zédé's *Gymnote*—*Gustave Zédé*—*Morse*—*Narval*—Two Spaniards—Russian submarines—The submarine in Germany

THE YANKEES TAKE THE PLUNGE   57
The *Intelligent Whale*—John P. Holland—*Holland No. I*—*Fenian Ram*—The Zalinski boat—Tuck and Baker—The *Plunger*—*Holland VI*—Simon Lake—*Argonaut Jr.*—*Argonaut* and *Protector*

"SUNK WITHOUT WARNING" 69
   Submarine development in the Royal Navy—The diesel engine—First U-boat—Submarines in World War I—Submarine developments between the wars

THE ENEMY IS BELOW 79
   The submarine and World War II—Antisubmarine warfare—Asdic—War against commerce—German submarine developments—Schnorkel—Walter's fast electric boat—Walter's hydrogen peroxide engine—Midget submarines—The frogmen

"UNDER WAY ON NUCLEAR POWER" 93
   Nuclear power—SINS—High speed, streamlined hulls—U.S.S. *Nautilus*—Under the North Pole—Ballistic missile submarines—Submarine cargo ships of the future

FREE AS A FISH 99
   Le Prieur's diving apparatus—Masks, flippers, and breathing tubes—Cousteau and the aqualung—Hunters and cameramen—"Rapture of the depths"—The deep divers—To explore the Continental Shelf—Sea houses—Sealab I and II

BALLOONS TO THE ABYSS 109
   Life in the depths?—Beebe and the bathysphere—Piccard's bathyscaphe—*F.N.R.S. 3*—*Trieste*—The deepest dive

UNDERWATER RESEARCH VESSELS 115
   Underwater research vehicles—*Deep Quest*—*Aluminaut* and *Alvin*—Atom bomb recovery—*Aluminaut* salvages *Alvin*—*Deep Diver*—Privately owned submarines

CHRONOLOGY 123

INDEX 127

THE PIONEERS

The urge to explore the cold wet world beneath the sea has been with us ever since some remote ancestor first thrust an arm into a clear pool to grab for a shellfish or crustacean. And we can be sure that it was not long after men braved the water and learned to swim that some daring soul took a deep breath and tried to see how far under the surface he could go. He found he couldn't stay down very long, but if the water was clear, he had a glimpse of a watery region far different and, in some ways, more fascinating than that on land, before he rose sputtering to the surface.

Thousands of years later men discovered wealth beneath the sea in the shape of the pearl oyster and the sponge. Those who sought these treasures were the world's first professional divers. Their equipment was simple—a good pair of lungs, a noseclip of wood or bone, a rope, and a stone weight. Later, crude attempts were made to provide an extra supply of air. Ancient Greek and Roman writers mention such efforts, but unfortunately do not go into details. However, a book by a Roman, Vegetius, who wrote on military matters about A.D. 375, shows a picture of a diver wearing a tight-fitting helmet—probably of leather—with a leather hose leading from it to the surface. The open upper end of this tube was fastened to a bladder, which kept it afloat. Such a device could only function a few feet below the surface. At any real depth the pressure of the water on the diver's chest would be too great for him to be able to breathe.

These primitive prototypes of diving gear were certainly a great rarity. For many centuries, when an individual dived, he did it in the old way, with no more air than he could carry in his lungs. (While pressure might be too great to allow the chest muscles to *expand* the lungs, a trained diver could *hold* his breath at a considerable depth.)

Such dives were necessarily brief. While men have gone down over 150 feet and stayed underwater for two minutes or more, holding their breath, a dive of 90 seconds to a depth of 60 feet is considered good. This means only a few seconds of actual working time on the bottom. But there was another way—a quite simple device—for men to go down for longer periods. This was the diving bell.

If you upend a glass and push it down into a container of water you will notice that a pocket of air is trapped against the bottom of the glass. The air pressure inside keeps the water from rising all the way. This is the principle of the diving bell or chamber, and if the "bell" is heavy enough and large enough, it can be lowered with one or two men in it. They breathe the trapped air

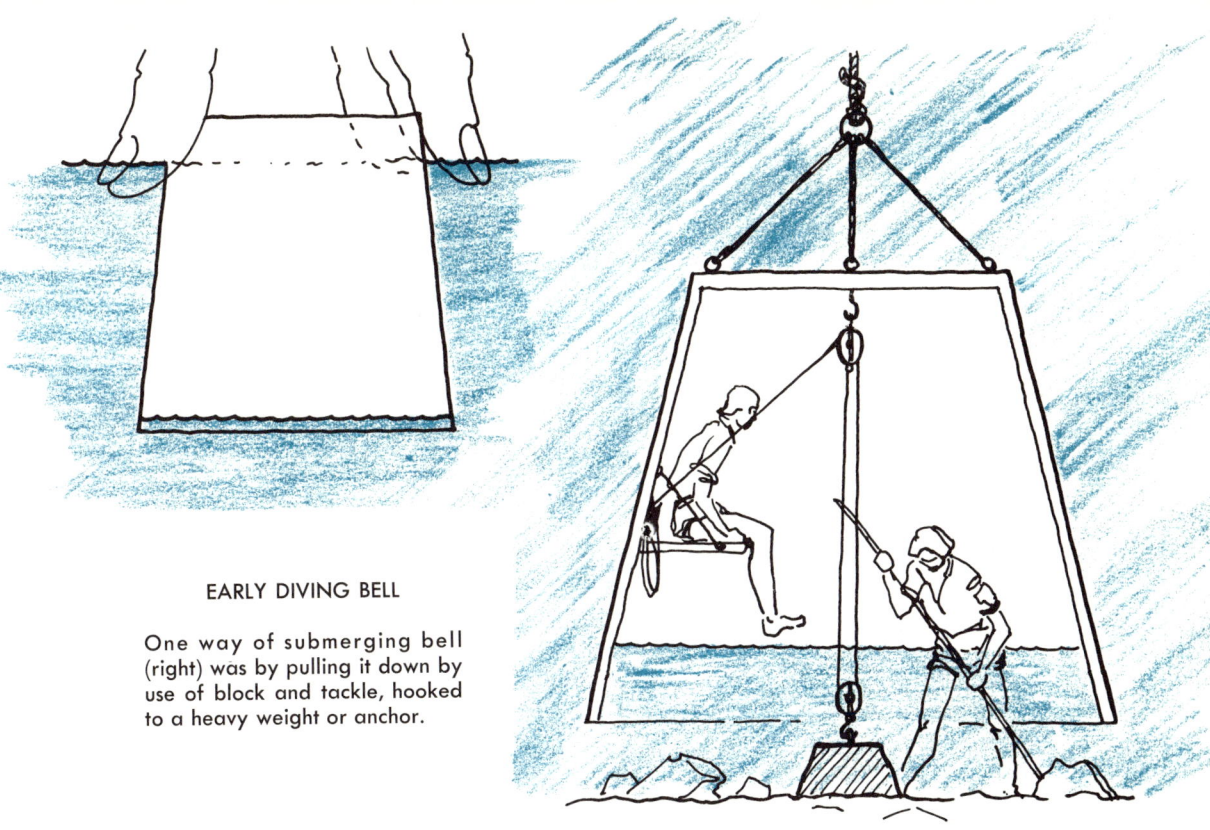

EARLY DIVING BELL

One way of submerging bell (right) was by pulling it down by use of block and tackle, hooked to a heavy weight or anchor.

and can remain down until it becomes too foul. Pressure of water increases with its depth, and at just over 30 feet the air is compressed so that the water rises about halfway up into the chamber. This meant that a bell of this type could only be used at comparatively shallow depths, but men could work on such things as foundations for sea walls or bridges.

No one knows when the first diving bell was used. There is a legend that Alexander the Great went down in some such device about 333 B.C. Some give credit for the invention to Roger Bacon in 1250. A bell was used to explore the wrecks of Roman galleys in Lake Nemi in 1535; others were used on various wrecks. One invented by the English astronomer Edmund Halley, with a primitive pumping device, operated at depths of 60 feet for as long as 90 minutes in 1690.

But the diving bell, while it did enable men to work beneath the surface for some time and with some degree of safety, was a relatively immobile affair—linked to the surface by ropes and chains. Men with imagination dreamed of a vessel of some sort which could navigate under water, free of any ties with the surface. It was a long time before anyone succeeded, but there were numerous attempts, and some of them were quite ingenious.

The difficulties were considerable. First, and most important, was the problem of constructing a hull which, with its entrance hatch, would be perfectly watertight. Second, there had to be some way to make the vessel not only submerge—this was all too easy, as a good many unfortunate people discovered—but also rise to the surface again when desired. And third, there was the question of how to propel the hull through the water. This was before the days of engines and screw propellers, and only human muscles were available.

In 1620, a Dutch physician named Cornelius Van Drebbel designed a craft built of wood, strengthened inside with iron bands, and covered with tightly stretched hides soaked in grease. Twelve oars provided the motive power, and the oarholes in the vessel's sides were made watertight by fitted leather sleeves. The vessel is said to have made a successful trip under the River Thames, but it is doubtful if the upper deck was more than awash. Possibly the oars might have forced her under a little for short stretches, and there was some sort of arrangement of valved tubes—one to take foul air out and the other to let fresh air in. This seems to have been an early ancestor of the schnorkel tube developed much later.

No plans of Van Drebbel's craft have been discovered, but there are drawings of a strange vessel built at Rotterdam in 1653 by a Frenchman named De Son. She was of wood strengthened with iron bands. (There was no way of rolling iron plates of any size in those days.) A solid beam, tipped with iron to act as a ram, ran through the center of the hull. The center section, or part of it, was open to the water and in it was a paddle wheel, revolved by clockwork. The hull could be ballasted until all but the small platform at the top was submerged.

Inventors in those days were given to exaggerating the capabilities of their brain children and De Son was no exception. His vessel, it was modestly claimed, could ". . . destroy [by ramming] a hondered Ships, can goe from Rotterdam to London and back againe in one day, & in 6 Weeks to goe to the East Indies. . . ." Actually, the 72-foot vessel wouldn't "goe" at all. The clockwork mechanism, which was strong enough to turn the paddle wheel while the boat was on the stocks, was not able to turn it against the resistance of the water. The inventor was reduced to exhibiting his strange vessel to the curious for a few pennies. But it was a good try, and De Son's design at least looked more like our idea of a submersible than some which came after.

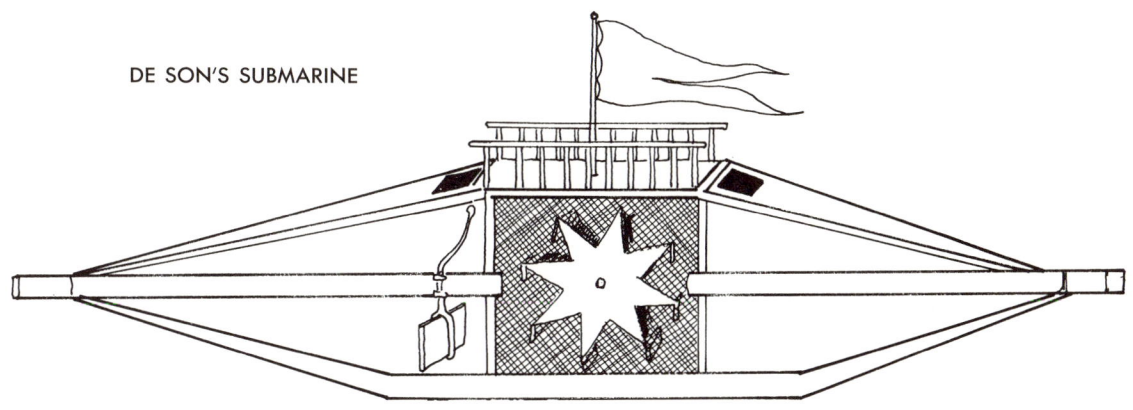

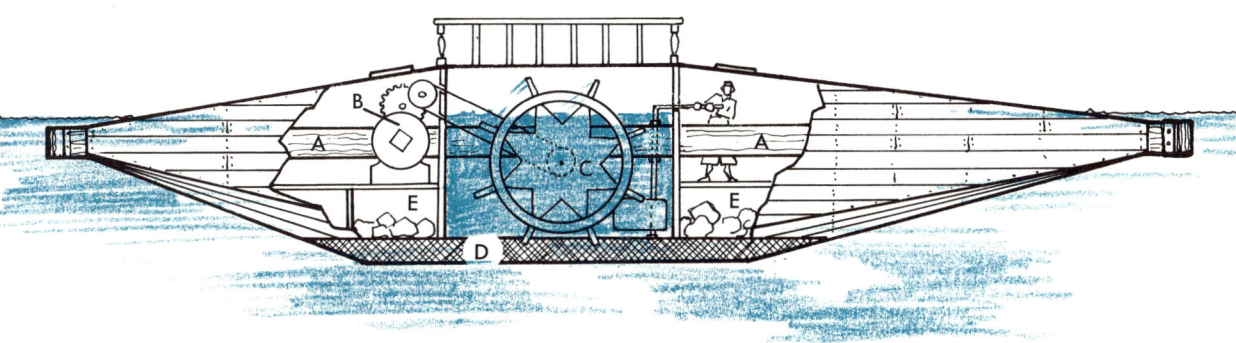

Above: De Son's submarine—1653—from an old print. Below: De Son's vessel as she may have actually looked.

        A — Central beam
        B — Clockwork mechanism for driving paddle wheel, C
        D — Keel
        E — Ballast

THIS IS DISPLACEMENT

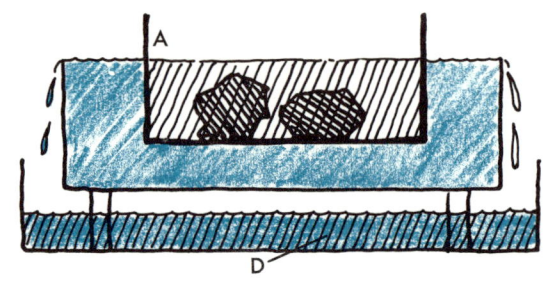

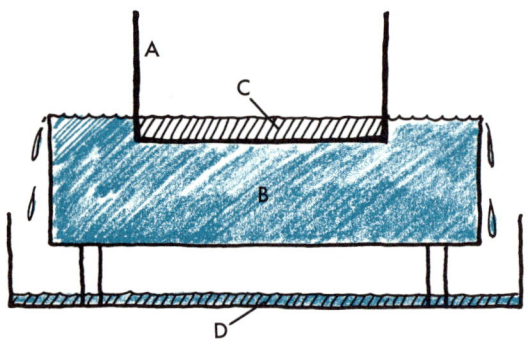

Left: Box, A, floats in tank, B. Volume of water displaced, C, is caught, D. Weight of water, D, equals weight of box, A. Above: Weight of D equals weight of box, A, plus weight of rocks.

Before taking a look at the development of underwater craft in the eighteenth century it might be well to understand a few of the terms used to describe the way a hull acts on and below the surface of the water, and how a hull may be made to submerge to a given depth and then to come up again. First, there is the matter of *displacement*. Surface vessels are designed to float with a portion of their hulls submerged. The submerged part takes up a certain amount of space (varying with the design and construction) and the weight of the water which would normally fill this space equals the weight of the hull, its engines, stores, crew, and everything in it. A simple example illustrates this. Suppose we put a light wooden box (watertight, of course) in a tub filled to the brim with water. Underneath we have an even larger tub. The wooden box will float, but not right on top of the surface. It will sink a little, and the water it displaces will spill over the edge of the tub into the tub below. If we weigh that water we will find that it weighs just the same as the box. This weight is called the displacement. A rock or two in the box will make it sink deeper, pushing more water out of the tub. And again, the amount displaced this time will equal the weight of the rocks.

Naturally, one cannot catch and weigh the water displaced by a real ship, but the volume of the part of the hull to be submerged can be estimated quite closely. This volume, expressed in cubic feet for a small vessel, is multiplied by 64, since a cubic foot of sea water weighs approximately 64 pounds. The figure arrived at represents pounds of displacement, and is the weight of the hull, fittings, motor, and fuel, plus ballast enough to sink the hull down to its designed water line.

When a hull is floating, the weight of the hull, plus cargo, is equal to the water displaced. The vessel is said to have *positive buoyancy*. But if the boat is overloaded, or if she springs a leak so that water (which is heavy) displaces the air (which is light) then the vessel loses its buoyancy. When the vessel's weight becomes more than the weight of the displaced water, the ship sinks. It has *negative buoyancy*.

A surfaced submarine has positive buoyancy. When it submerges, air is let out—vented—of areas of the hull, called the ballast tanks, and water is admitted. As the air escapes out of the vent valves at the top of the ballast tanks, water rushes in through flood ports in their bottoms, and the vessel, losing buoyancy, sinks beneath the surface.

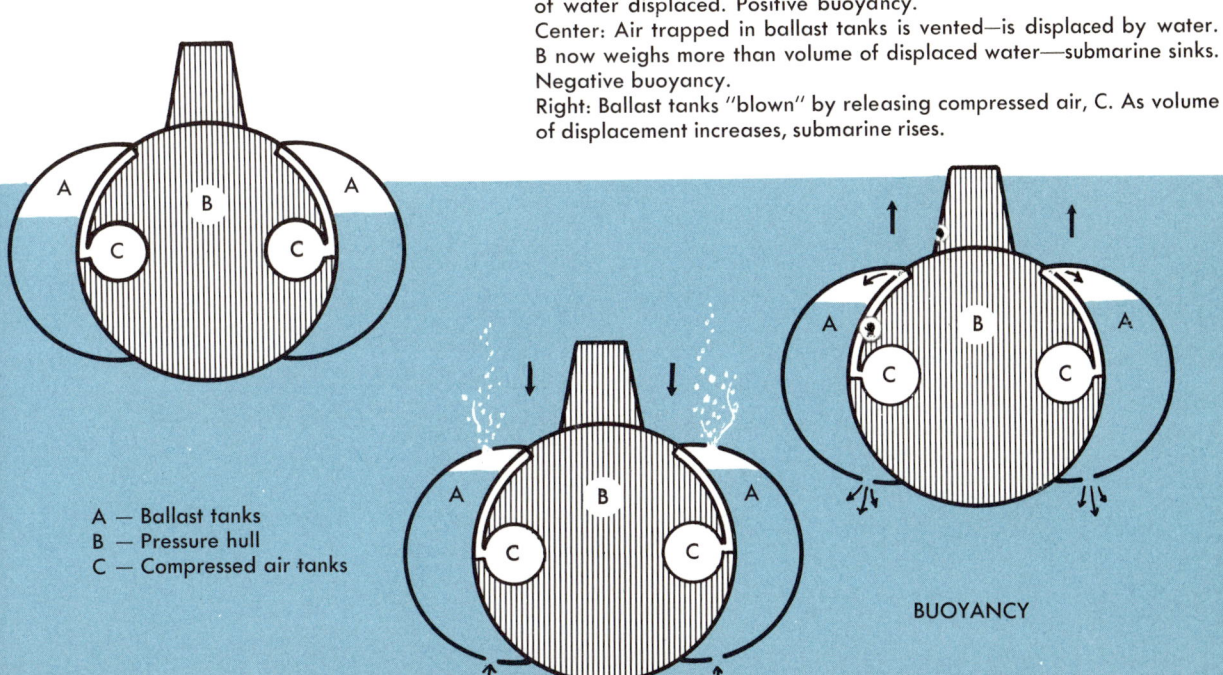

Left: Submarine on surface. Weight of hull and contents equals weight of water displaced. Positive buoyancy.
Center: Air trapped in ballast tanks is vented—is displaced by water. B now weighs more than volume of displaced water—submarine sinks. Negative buoyancy.
Right: Ballast tanks "blown" by releasing compressed air, C. As volume of displacement increases, submarine rises.

A — Ballast tanks
B — Pressure hull
C — Compressed air tanks

BUOYANCY

Naturally, if this state of negative buoyancy was allowed to go unchecked, the submarine would continue to sink. Water pressure increases at the rate of 14.7 pounds per square inch (one atmosphere, so-called because that is the normal pressure of air at sea level) for every 33 feet of depth. Although modern submarines are built to take great pressure, below a certain depth even the strongest would be crushed. When pressure gauges inside the vessel show that it has reached the desired depth, air under high pressure is valved into the ballast tanks and some of the water is forced out. When just the right amount is blown out, the vessel is in a state of neutral buoyancy, weighing neither more nor less than the water it displaces at that depth.

The submarine thus can cruise at any desired depth within its safety limits—that is, at depths at which the hull is not in danger of being crushed. But there are several factors involved and keeping the vessel at its desired depth is a very tricky business. For one thing, water itself is heavier the deeper down you go. The weight of the water above actually compresses that beneath, making it denser and, thus, heavier. The deeper a submarine commander wishes to go, the more water ballast he must let into his ballast tanks. Because cold water is denser than warm water, it takes more water ballast to submerge to a given depth in cold water than it does in warm. Because the ocean is crisscrossed with submerged currents of different depths and temperatures, it will be seen that depth-keeping is a complex operation.

This is the basic principle behind the art of the submariner—but in the days of De Son and Van Drebbel this knowledge was far in the future. All the pioneers of undersea travel had were some theories, a great desire to be able to navigate beneath the seas—and a large share of courage.

BARRELS AND BELLS

By the beginning of the eighteenth century, scientists had learned something about pressure and its effects. It was plain that man could not breathe normally at any but the shallowest depths, even with his head in a watertight helmet connected to the surface with an air tube. The obvious thing to do was to put him in some kind of armor. A successful device using this idea was invented by an Englishman named John Lethbridge in 1715.

Lethbridge's "suit" consisted of a cask, hooped with iron "without and within," a glass porthole some 4 inches in diameter and 1¼" thick, and two holes, with leather sleeves attached, for his arms. After climbing into his cask, the end was bolted on and the diver was lowered into the water. The air supply was sufficient for about thirty minutes, then he was hauled up and his assistants unscrewed a couple of air holes, and renewed the air with a pair of bellows. It was not the most mobile of "engines," but Lethbridge made many dives and, nearly twenty years after his first descent, was diving for money lost on a ship wrecked in Marseilles Harbor. He claims he often worked at 10 fathoms (60 feet).

About 1772, a Frenchman named Freminet discarded the "man in a barrel" idea and made a diving dress with a helmet attached to a watertight leather suit. A jointed metal frame supported the leather against the pressure and allowed the diver some freedom of movement and a chance to breathe more or less normally. The helmet was too small to contain much air, so a leather breathing tube was used. A spring-operated bellows fixed to the suit helped circulate the air. With such a suit, dives of an hour's duration at 50 feet were made, and useful work performed underwater.

There are various mentions of underwater vessels in the early eighteenth century and several patents were taken out. Most of these were just ideas, the brain waves of some scholarly types which never got beyond the planning stage. But in 1747, an Englishman named Symons actually built a boat designed to be submerged until her upper hull was awash, if not actually under water. A contemporary drawing shows a normal boat's hull but decked over with a domed superstructure. The whole hull was covered with greased leather to render it watertight. In the bottom of the hull were fastened several leather sacks, with their bottoms open to the water. When the necks of these sacks were untied they filled with water and the weight of the hull and ballast sank the boat until she was awash. To rise, the water was squeezed out of the sacks by a man heaving on a lever, and the necks tied

EARLY BALLAST TANKS

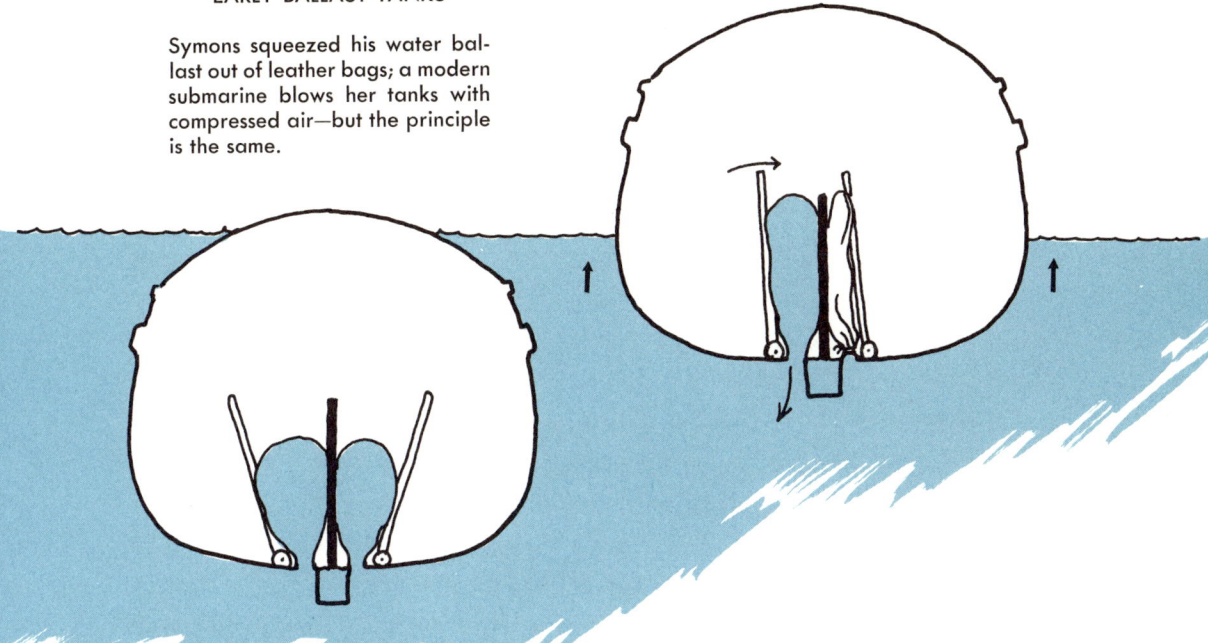

Symons squeezed his water ballast out of leather bags; a modern submarine blows her tanks with compressed air—but the principle is the same.

to prevent the water from entering again. Although very primitive—and how well it really worked, we don't know—the idea is the same as is used in modern submarines, with steel ballast tanks replacing the leather sacks, and water forced out by compressed air instead of levers.

In 1773, an English ship's carpenter named Day built a small boat in which he descended some 30 feet. Day pioneered the use of detachable ballast—in this case, large stones hanging beneath the vessel which could be let go from inside the hull. Thus lightened, the vessel rose to the surface. The success of Day's first boat led to the construction—or rather, conversion—of a larger vessel. A submergence was made in the new vessel in Plymouth Harbor and a second was made in a deeper area where the soundings showed 22 fathoms. There was difficulty making the hull submerge and Day ordered more weights added. The craft then went down to the bottom, but the pressure at 132 feet is almost 75 pounds per square inch, and this was undoubtedly too much for the vessel's hull. Whatever the cause, she never came up, and Day is listed among the early victims of the fight to conquer the world under the sea.

Not long after Day's last and fatal dive, a young man on the other side of the Atlantic planned and built the most ingenious underwater craft yet seen—one which not only worked, but which very nearly succeeded in fulfilling its purpose, which was to sink an enemy warship. The man was David Bushnell and his justly famous *Turtle* was the ancestor of a long line of American underwater craft.

Bushnell was the inventive-minded son of a Connecticut farmer. Largely self-taught, he had entered Yale at the age of thirty-one. Among other things, Bushnell was intrigued with the idea of exploding gunpowder underwater. The intense feeling which was shortly to find expression in the American Revolution gave a practical turn to Bushnell's experiments. But the powder-filled mine with its clockwork timing device was the least of Bushnell's inventions. A mine was of little use without a means of placing it against an enemy hull. So, on the Connecticut man's drawing board, there grew plans for a one-man submarine—a vessel which would take its mine to the enemy, fix it to its victim, and retire unseen.

By this time the first battles of the Revolution had been fought and the blockade of rebellious ports had begun by the Royal Navy. If ever there was need for a device which could strike deadly and invisible blows at the enemy fleet, it was now. Bushnell left Yale and retired to his home near Saybrook to build his submarine.

Two turtle shells clamped together best describe the shape of the hull, so the name was appropriate. The "shells" were made of oak, reinforced inside against pressure and bound with iron hoops. On top was a low, round, flat-topped structure—not much bigger than a lady's hatbox. This conning tower had glass ports so that the operator could see out, and the top formed a hatch, firmly bolted in place when the "crew" embarked.

The whole contraption was far ahead of anything that had been designed up to that time. Propeller, depth gauge, snorkel tubes, and ballast pumps and tanks remind us more of the nineteenth than the eighteenth century. Fox fire (the phosphorescent material

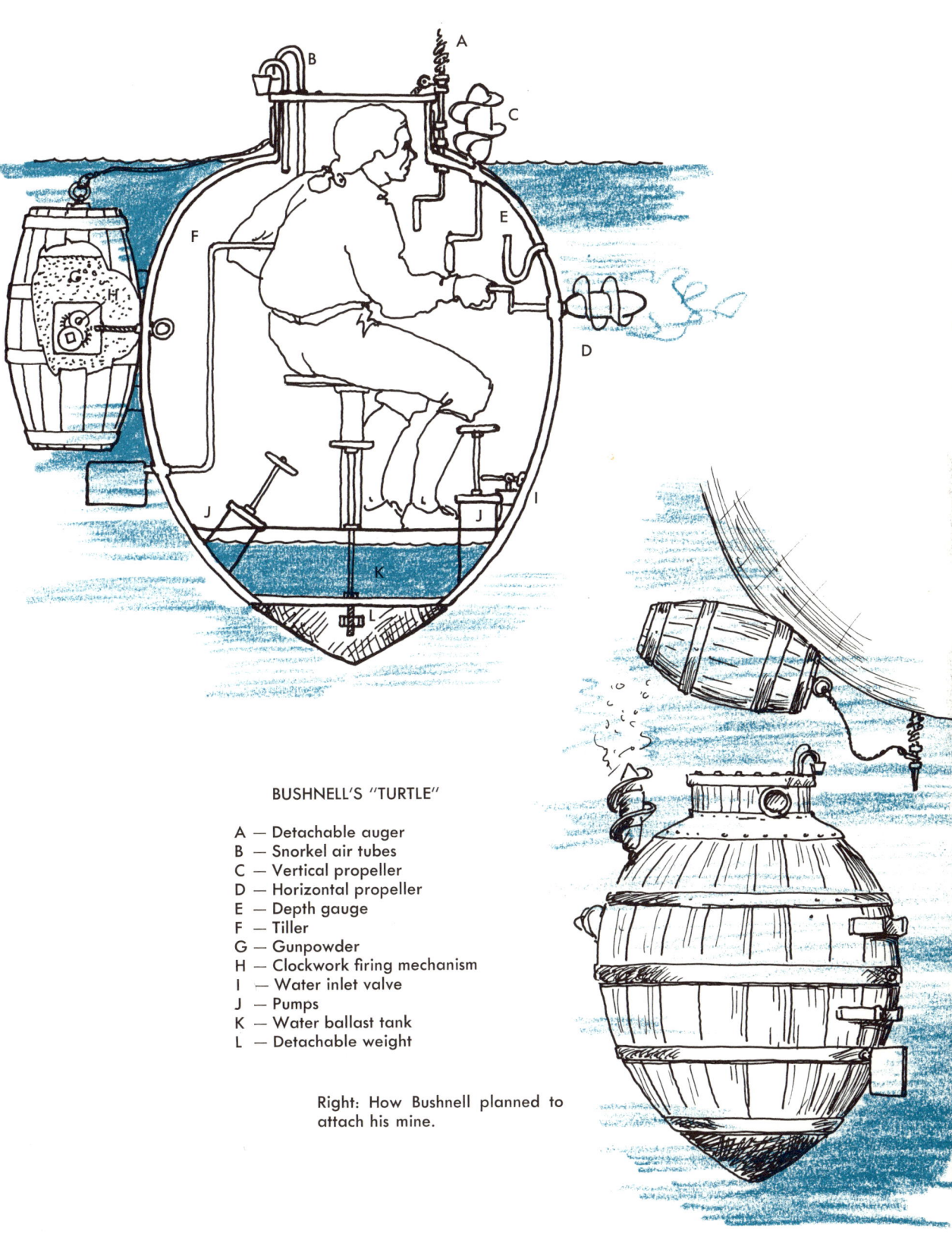

BUSHNELL'S "TURTLE"

A — Detachable auger
B — Snorkel air tubes
C — Vertical propeller
D — Horizontal propeller
E — Depth gauge
F — Tiller
G — Gunpowder
H — Clockwork firing mechanism
I — Water inlet valve
J — Pumps
K — Water ballast tank
L — Detachable weight

Right: How Bushnell planned to attach his mine.

sometimes found on decaying timber) was used to mark a compass. A cork, floating on top of the water column of the pressure depth gauge, was tipped with the material so that the depth could be read in the darkness. As the air inside the hull was only sufficient for thirty minutes, the approach to the target was to be made at night, with the conning tower and snorkel breathing tubes just out of water. Then, when the target was reached, enough water ballast was admitted to barely submerge the hull, which was then forced down by the vertical propeller. Once under the enemy vessel, the detachable auger bit was to be screwed into the bottom of the enemy ship and the bolt holding the casklike mine to the submarine would be withdrawn. Withdrawing this bolt started a clockwork device which, in a given time, activated a flintlock mechanism, which in turn set off the charge of gunpowder.

The submarine was ready in the autumn of 1776 and it was intended to use it against the British shipping in New York Harbor. The specific target for the first attack was to be the British flagship—H.M.S. *Eagle*. Then Lady Luck abandoned Bushnell. He was in poor health, and to crank the *Turtle's* screws, steer, and work the ballast pumps called for considerable physical stamina. His brother, who had tested the *Turtle* and thoroughly understood its operation, was to crew the boat. Then the brother fell ill. The last-minute substitute—Sergeant Ezra Lee—actually made his way underneath *Eagle*, but the auger bit refused to go into the hull. Probably he had hit on the one place where the heavy iron strap connected the rudder fitting to the sternpost. The submarine bobbed up to the surface—still unobserved—but daylight was approaching and Lee decided not to try again. On the way back to his launching point on Manhattan he released the mine, which duly exploded, causing much alarm to the invading squadron. It was a good try, and had it succeeded it would have undoubtedly advanced the development of undersea vessels by half a century.

Not long after Bushnell had designed the *Turtle*, a noted British engineer, John Smeaton—builder of the Eddystone lighthouse—

built a workable air pump and used it to force air down into a diving bell. A reservoir of air provided temporary relief in case the pump broke down, and the air hose was provided with a non-return valve so that the air forced down stayed where it belonged —in the bell. Smeaton's bells were used on many underwater engineering projects—laying foundations, repairing piles, and constructing sea walls, quays, and canal locks. Modern diving bells are similar in principle, with the added refinement of telephonic communication, electric cables for power, and air hoses for pneumatic drills or riveters.

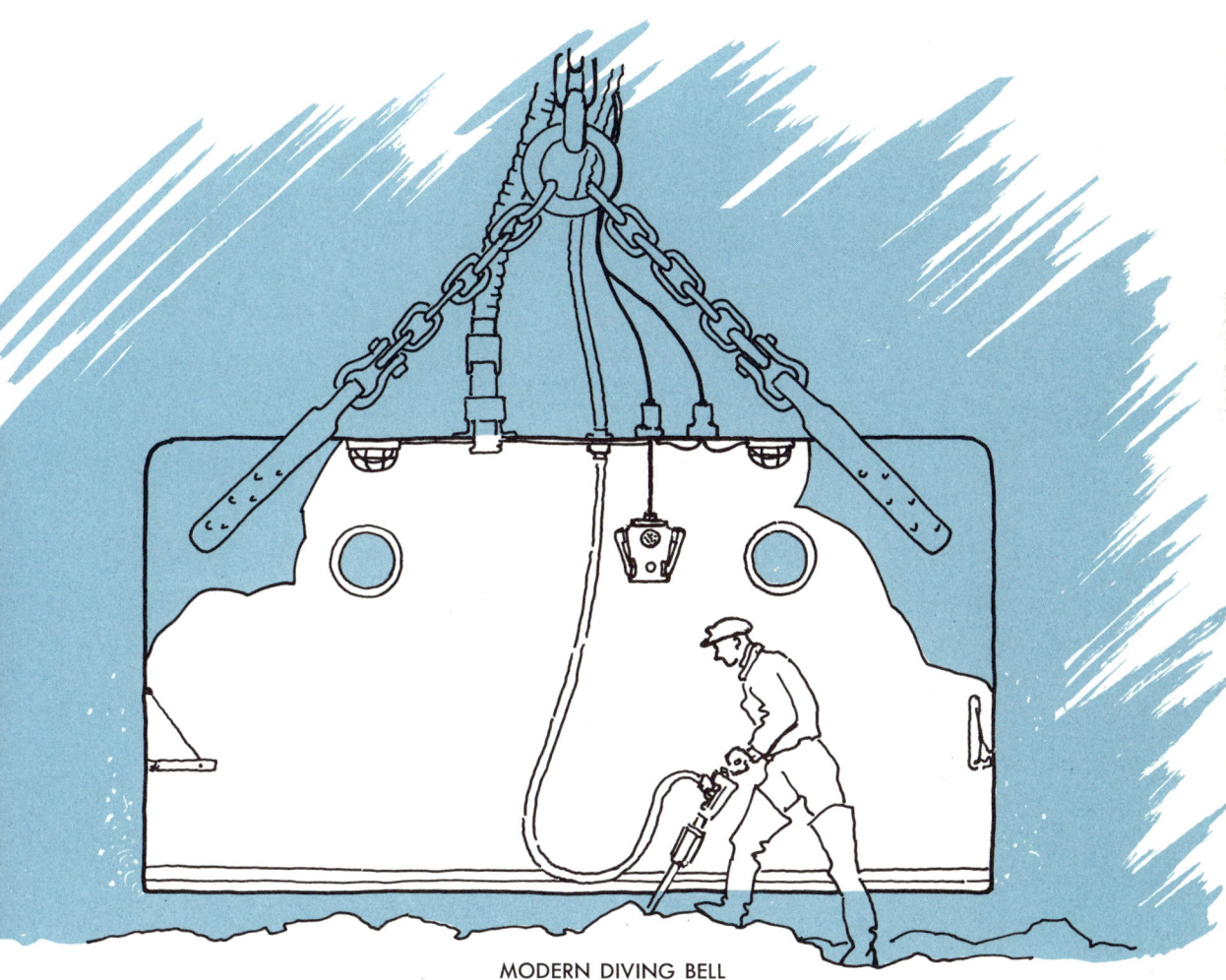

MODERN DIVING BELL

FULTON'S "NAUTILUS"—1801

Length: 21' 4"   Diameter: 7'

A — Conning tower
B — Rod with detachable spike, C (shown in enemy hull)
D — Towed mine, with clockwork or trigger activated firing mechanism
E — Collapsible mast, with operating gear, F
G — Anchor and windlass
H — Crank for turning propeller
I — Gearing for vertical (J) and horizontal (K) rudders
L — Pumps
M — Ballast tanks

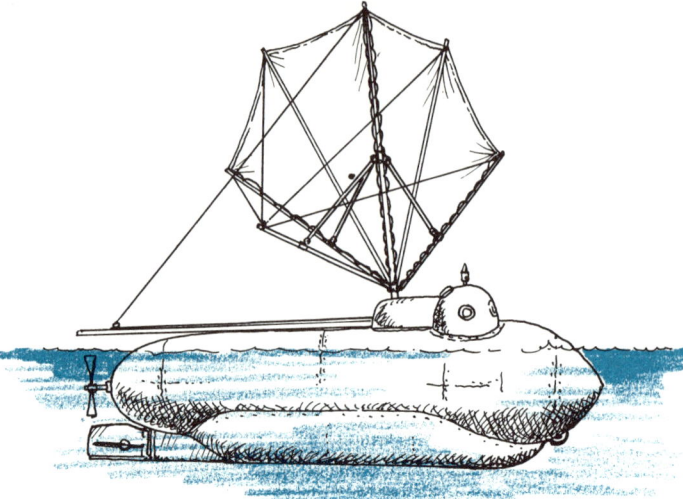

"Nautilus" on surface with sail set

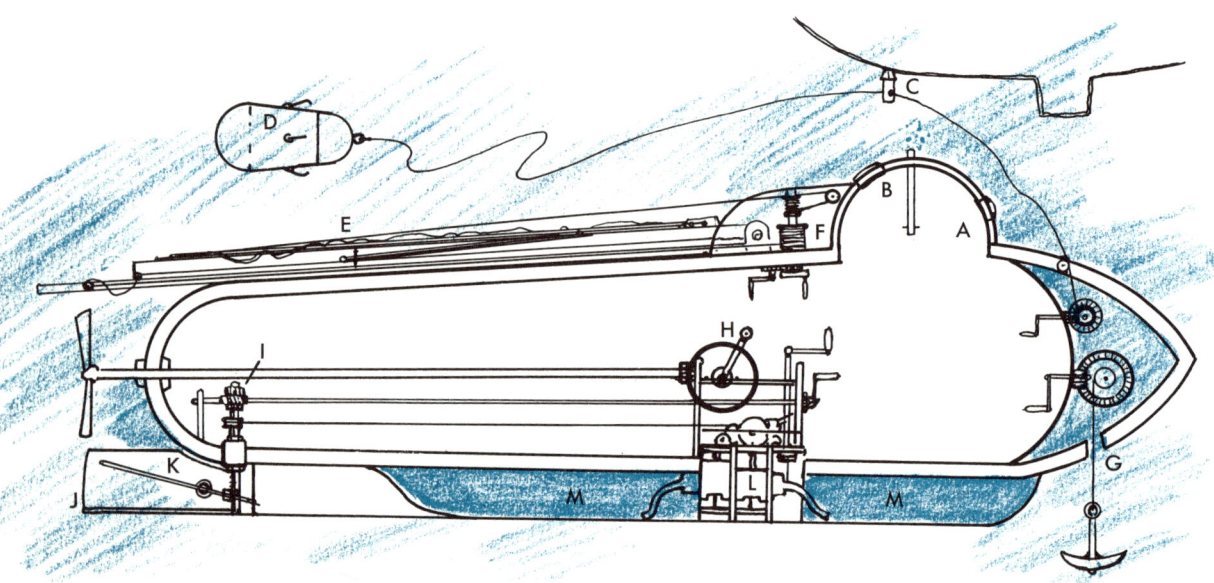

Toward the end of the eighteenth century another American, Robert Fulton, also became interested in the possibility of an undersea boat as a means of bringing an explosive device in contact with the hull of an enemy ship. Fulton was undoubtedly a genius—jeweler, artist, engineer, designer of ships and machinery. His submarine, planned in France and offered to the French Directory, was laid down in late 1800 and launched in May, 1801. The hull of the *Nautilus*, as she was named, was of copper plating over an iron frame. Ballast tanks admitted water to decrease buoyancy and a heavy iron keel—detachable in case of emergency—was fitted along the bottom. There was a dome-shaped conning tower with view ports. From the tower a spike could be driven to anchor a clockwork-timed mine. A hand-operated propeller drove the hull forward, and when in a state of neutral buoyancy, two inclined planes, acting as diving rudders, regulated the vessel's depth.

There was a sail on a collapsible mast for use when on the surface, and double cylinder pumps were installed to force the water out of the ballast tanks when ascending. With a crew of three aboard, the craft dived to a depth of 7 meters (about 25 feet) in the harbor of Brest and stayed below for an hour. Later a reservoir of compressed air extended this period considerably. Towing a charge of powder in a cask, with a firing device in it, Fulton succeeded in blowing up an old hulk put at his disposal as a target by French officials. In her best run the vessel covered 500 yards in seven minutes and returned to her starting point.

But the French officials were not prepared to sanction the use of such a terrible invention—although General Washington had given Bushnell his blessing. It may have been conscience, the honor of the French Navy, or Fulton's demand that his crew should be granted the status of belligerents (so that they could not be hanged as pirates if captured). In any event, his invention was turned down. Having failed in interesting the French in blowing up British vessels, Fulton tried to interest the British in similar means of destroying Napoleon's ships. Many tests were made, but the reaction of the British investigating board was much the same as that of the French. The veteran Admiral Earl St. Vincent, First Lord of the Admiralty, saw the possibilities, but also saw that such a weapon favored the nation with the inferior fleet. England was already the leading naval power and he wanted no part in the development of any device which might threaten that leadership.

Before Fulton returned to America and steamship fame, the British offered him money to suppress his invention, but he would only accept enough to cover his expenses. He appears to have been genuinely interested in world peace and free trade among nations, and had hoped that the acceptance of his device might have made war—at least naval war—impossible. In later years, governments grew less squeamish and a hundred years after Fulton's *Nautilus*, navies of all nations were feverishly competing to design the deadliest undersea warships.

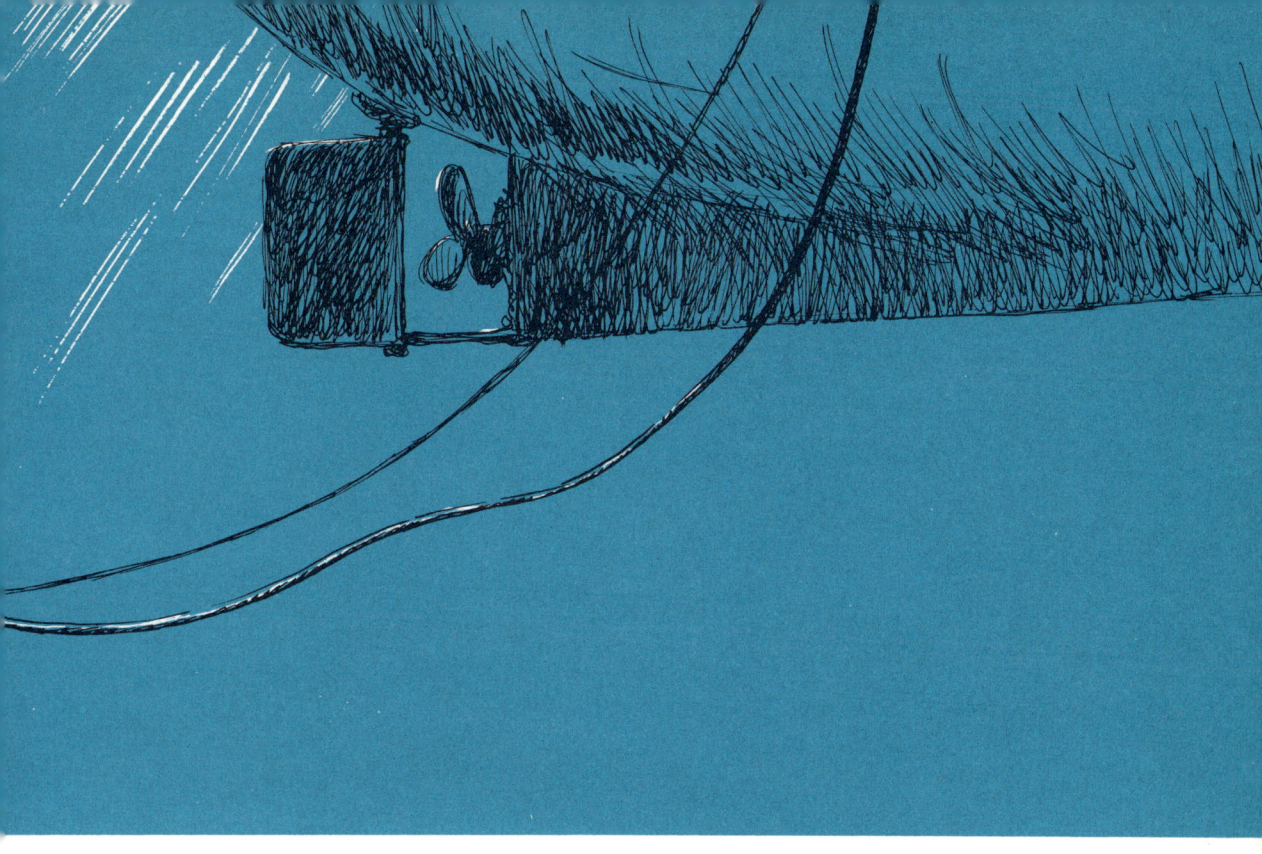

THEY WALKED UNDER THE WATER

The development of efficient apparatus for the individual diver went on meanwhile and, until well into the twentieth century, it was devoted almost exclusively to the peaceful pursuit of underwater construction, engineering, and salvage. The diving bell had been more or less perfected, but such devices, useful as they were in many instances, were too cumbersome and lacking in maneuverability to allow their occupants to perform many necessary operations. The next step was to apply the principle of the diving bell, with its air forced down by pumps, to the diving suit. Make the bell small enough to rest on a man's shoulders—and we have a modern diving helmet. This bell-helmet was invented by a naturalized Englishman named Augustus Siebe in 1819. The helmet,

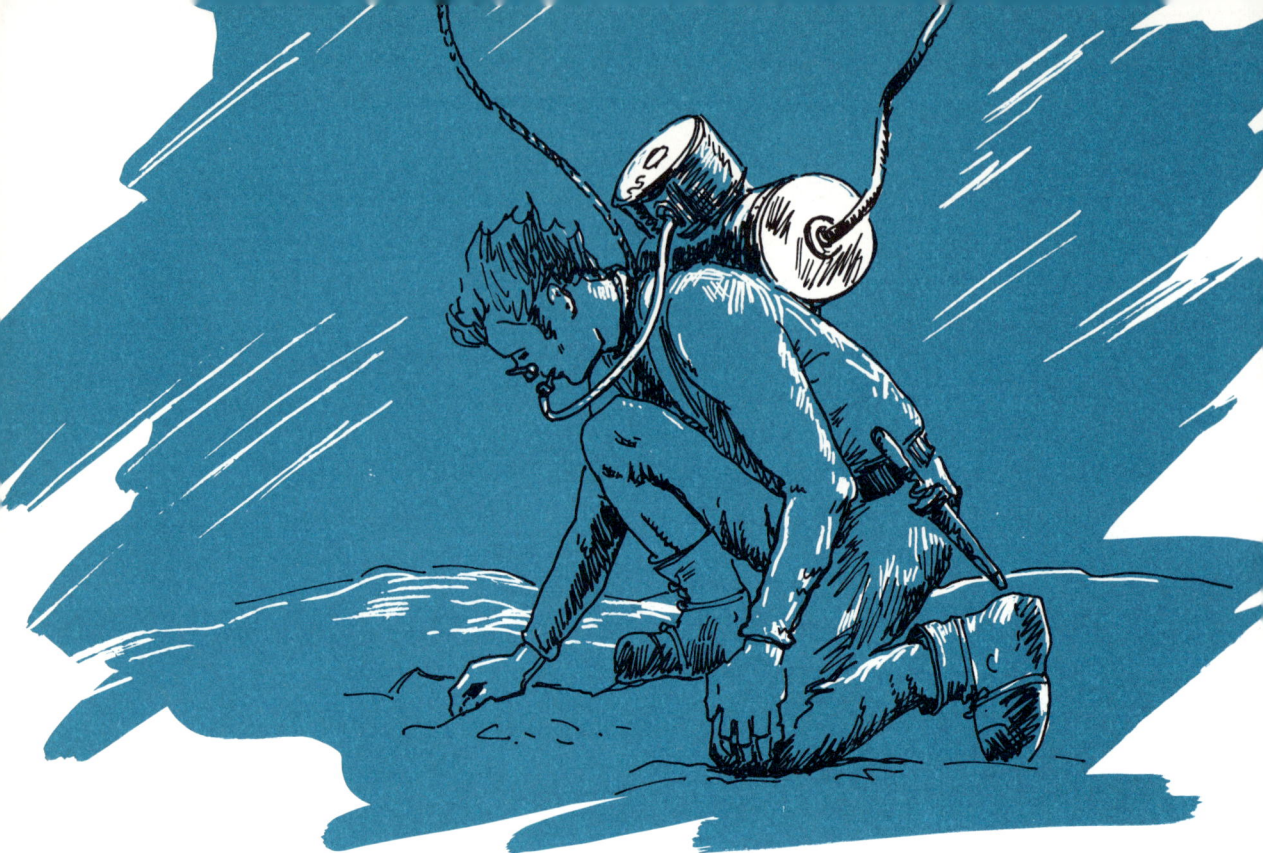

ROUQUAYROL-DENAYROUZE APPARATUS—c. 1860

with a faceplate, was fastened to a waterproof leather jacket. The air was under sufficient pressure to keep the water in the suit below the wearer's chin. The excess escaped around the bottom of the suit. Weights on the diver's feet took him down and helped keep him upright.

Siebe's helmet was a success but it had its drawbacks, one of the chief being that if the diver bent over too far, or stumbled, the water filled his suit. If he was not hauled up quickly, he probably drowned. Despite this, the helmet was much used, but Siebe kept on experimenting and in 1837 the design for a complete suit of waterproof canvas, with nonreturn air inlet and buoyancy regulating outlet valves, was patented. This suit, with improvements, is the type used today. Its introduction opened up vast new fields of underwater endeavor, giving its wearers comparatively great mobility at depths hitherto out of reach.

In 1840, using Siebe's suits, a team of British military engineers carried out the first organized salvage operation and founded the first school for naval divers. Their goal was to raise the valuable brass cannon from the wreck of H.M.S. *Royal George*, sunk accidentally at her moorings at Spithead in 1782 with some 1,000 persons aboard. The wreck rested, covered with silt and weeds, in 165 feet of water. The operation lasted several years and thousands of dives were made. The metal was brought up and the hulk systematically destroyed with underwater charges, electrically detonated.

An important invention was made during the years 1860-65 by two Frenchmen, Benôit Rouquayrol and Auguste Denayrouze. In their apparatus a metal cylinder containing air under considerable pressure was strapped to the diver's back. A tube led from the tank to pumps on the surface; another tube was held in the diver's mouth. A noseclip was used, but no helmet. A "regulator" valve, between the cylinder and the diver's mouth, automatically passed air at the required pressure, equaling that of the water at the depth the diver was in and thus facilitating breathing. An added advantage was that, for a few minutes, the diver could disconnect the surface air hose and operate freely on the air stored in his pressure tank. The apparatus was never popular with divers, possibly because they found the noseclip and mouthpiece inconvenient. But eighty years later the "regulator" idea, much refined, was to be used by Cousteau and Gagnon in the now famous aqualung.

The helmet-suit diver was tied to the surface by his air hose. Once freed from this encumbrance, the diver would be able to move freely and perform tasks impossible with the regular suit. In 1879, a British merchant marine officer, Henry Fleuss, invented the first fully independent diving apparatus. A watertight mask fitted over the face. Two breathing tubes ran from the mask to a bag worn on the diver's back. This in turn was connected to a tank holding compressed oxygen. The bag contained an absorbing agent (rope yarns soaked in a solution of caustic potash) which removed the carbon dioxide from the diver's exhalations.

Fleuss took his idea to Siebe's diving equipment company. It was used successfully in the 1880's in the then famous Severn Tunnel. The tunnel had flooded accidentally, and before it could be pumped out, a watertight iron door inside the tunnel had to be closed. This meant going down a 200-foot vertical shaft and then 1,000 feet along the flooded tunnel to the door. It was impossible for a diver to drag his hose and lines that distance, but one wearing Fleuss' rebreathing apparatus managed it.

Later, Fleuss and Robert Davis, an engineer of Siebe, Gorman & Co., improved the device. In 1911, Davis patented his submarine escape apparatus, basically a face mask with rebreather tubes and an oxygen air tank. Other inventors devised somewhat similar types of closed circuit apparatus, and in addition to providing equipment used by firemen, mountain climbers, and aviators, these rebreathing devices were used by "frogmen" and underwater warriors in World War II.

DAVIS ESCAPE APPARATUS

(A) Man, wearing mask and oxygen rebreather bag, enters escape lock. (B) Door is closed and water is admitted to lock, compressing air. When this equals water pressure outside, hatch is opened and man rises to surface. (C) Hatch is then closed by gears and water blown out by compressed air.

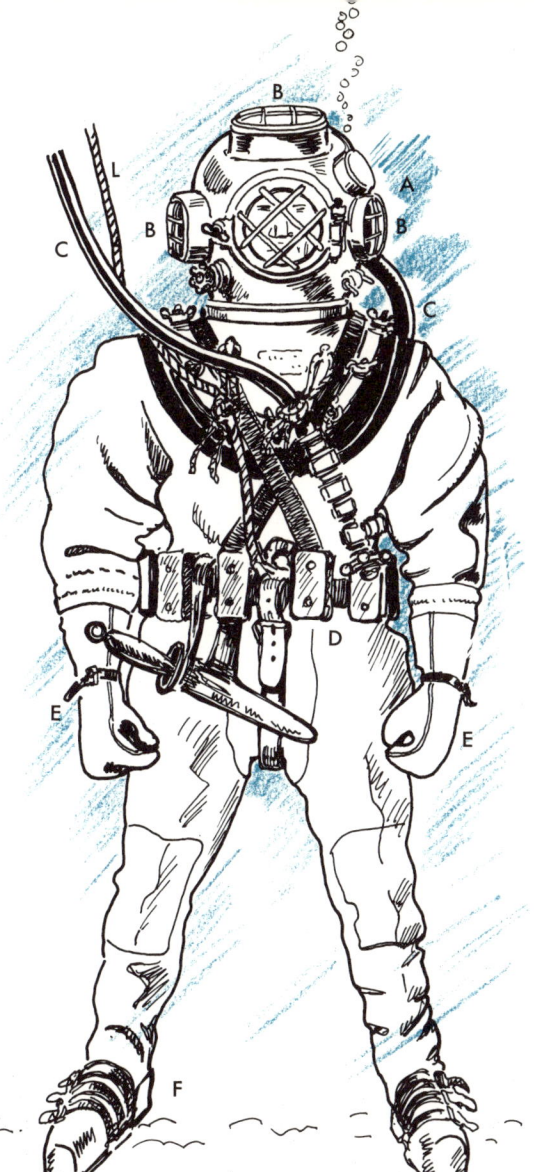

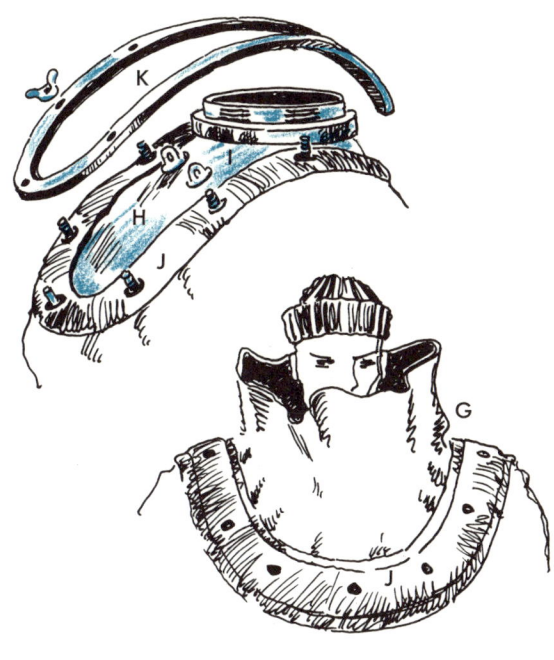

MODERN DIVING DRESS

A — Metal helmet with side and top view ports
B — Hinged faceplate and air release valve
C — Air hose with valve
D — Belt holding 5-pound removable lead weights and knife
E — Rubber mittens clamped to sleeves
F — Shoes with lead soles may weigh 35 pounds
G — Opening at top of suit wide enough to admit diver, feet first. Then metal breastplate, H, with collar ring, I, is slipped under edges of suit, J, and clamped in place with metal clamping ring, K. Helmet is then screwed down on ring, I.
L — Lifeline

The oxygen rebreather type of outfit was not intended for prolonged work at great depth. It had its uses, but the conventional diving suit was, and still is, in common use for the majority of underwater work. So, with the invention of Siebe's helmet suit, we can leave the "hard-hat" diver, with his rubberized canvas suit, bronze helmet, and clumsy weighted boots. Telephones have been added, and cables for floodlights, but essentially the helmet suit has not changed for 130 years.

But while the helmet diver's equipment has changed very little over the years, scientists have learned a great deal about pressure and its effect on man. Actually, although divers had frequently complained of curious ailments—aches and pains and cramps—it was tunnel workers and their ills which sparked the scientific research.

Like the helmeted suit, the steel or iron tube called a *caisson* was first developed around 1840. Used extensively for underwater tunnels and foundations, the caisson—the type using compressed air—is essentially a large tube made of a series of connecting iron rings. As the underwater excavation progresses, more rings are added until the tunnel is complete. To prevent water filling the section in which the actual excavating is being done, air is forced into the working chamber—which is entered through an air lock—

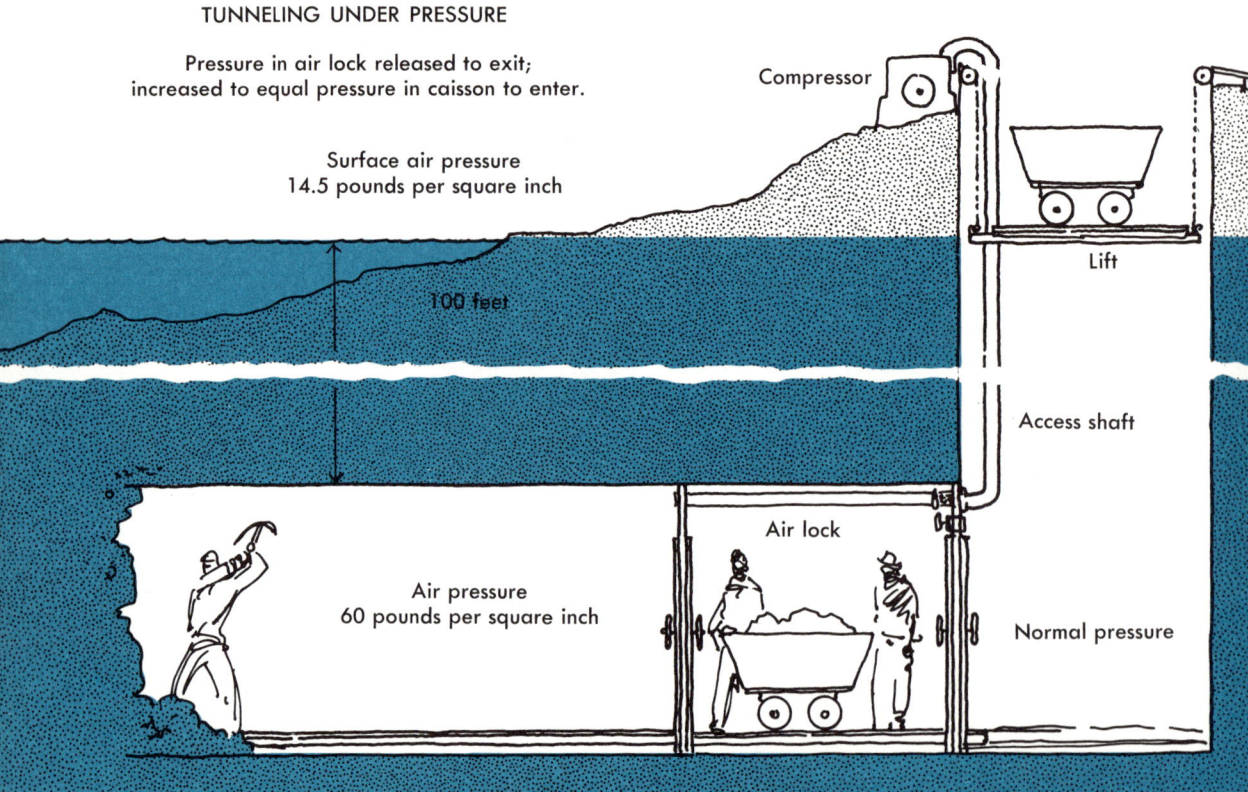

TUNNELING UNDER PRESSURE

under sufficient pressure to keep the water out. At a depth of 100 feet this means an air pressure of some 60 pounds per square inch. It was noticed that workers employed in these pressurized caissons suffered from pains in the joints and muscles, partial paralysis, and sometimes death. In America, during work on the foundations for a bridge at St. Louis (at 4½ atmospheres), of 600 workmen, 119 were affected and 14 died. None of those affected by what came to be known as "caisson disease" or "the bends," suffered while under pressure or while pressure was being raised. It was only after leaving the pressure chamber that ill effects became apparent.

It was finally discovered that, in breathing for a prolonged period of time under pressure, some of the nitrogen in the air is absorbed into the blood stream. Then, if the pressure is suddenly released, the nitrogen in the blood bubbles off—much like soda water when the pressure is released by removing the bottle top. These bubbles particularly affect the muscles and joints and can be exceedingly painful. In serious cases permanent injury or death can result.

Once the ill effects of too rapid decompression were definitely proved, tables were worked out showing the exact amount of time a diver working at a certain depth for a given time should take in ascending to the surface. Down to 33 feet, no decompression is necessary, since nitrogen bubbles are unlikely to form at only twice surface pressure. But a diver who had descended and worked for, say, 60 minutes at a depth of 100 feet, should take just over half an hour to ascend. This is usually done in stages—a 5-minute stop at 30 feet, 10 minutes at 20 feet, and 15 minutes at 10 feet. Had the diver remained only 10 minutes on the bottom, a 7-minute ascent would be sufficient. Dives to 200 feet or more must be of short duration. At 230 feet a diver spends only about 10 minutes (including his descent) before ascending, which takes 40 minutes in six stages. Recently, combinations of various gases—helium is one—have been used instead of compressed air at great depths. As a result, divers can now go deeper and with greater safety.

Should an accident force a diver to ascend too rapidly—a malfunctioning air escape valve could fill his suit like a balloon and shoot him to the surface like a cork—the diver is either sent down again immediately, to ascend slowly in proper stages, or is put into a steel tank, called a decompression chamber. Air is pumped in to raise the pressure to correspond to the depth at which the man had been working and then slowly released. These chambers have saved many divers from serious injury and are standard equipment on many salvage vessels and diving tenders.

Obviously, this slow decompression is a time- and money-wasting process. A diver who spends an hour below at 160 feet should take just over two and one-quarters hours in decompression time. This is one of the reasons for the present-day experiments with Sealabs and other pressurized "homes" beneath the sea. But that story could only come when divers had been freed from their cumbersome bronze helmets, weighted belts and boots, and entangling air hoses and life lines. First, let's turn to the development of underwater vessels. Designed primarily for war, they were vessels which, in the following century, progressed from boxlike iron freaks, scorned by naval authorities, to the deadliest craft known to man.

THE DEADLY FISH

In the minds of the inventors who dreamed of successfully maneuvering a vessel underwater, there was only one possible use for it. There was no point in designing an undersea craft to carry cargo. Surface ships could do that perfectly well. But a warship—which might in some way be made to submerge and approach an enemy unseen—that was a different matter. It was a frightening thought, one likely to give imaginative naval officers nightmares.

So it was that almost every war involving a sea power brought forth a crop of ideas by which the weaker fleet might destroy the vessels of the stronger. In 1850, Denmark and Germany (not then the great power which in the twentieth century twice plunged the

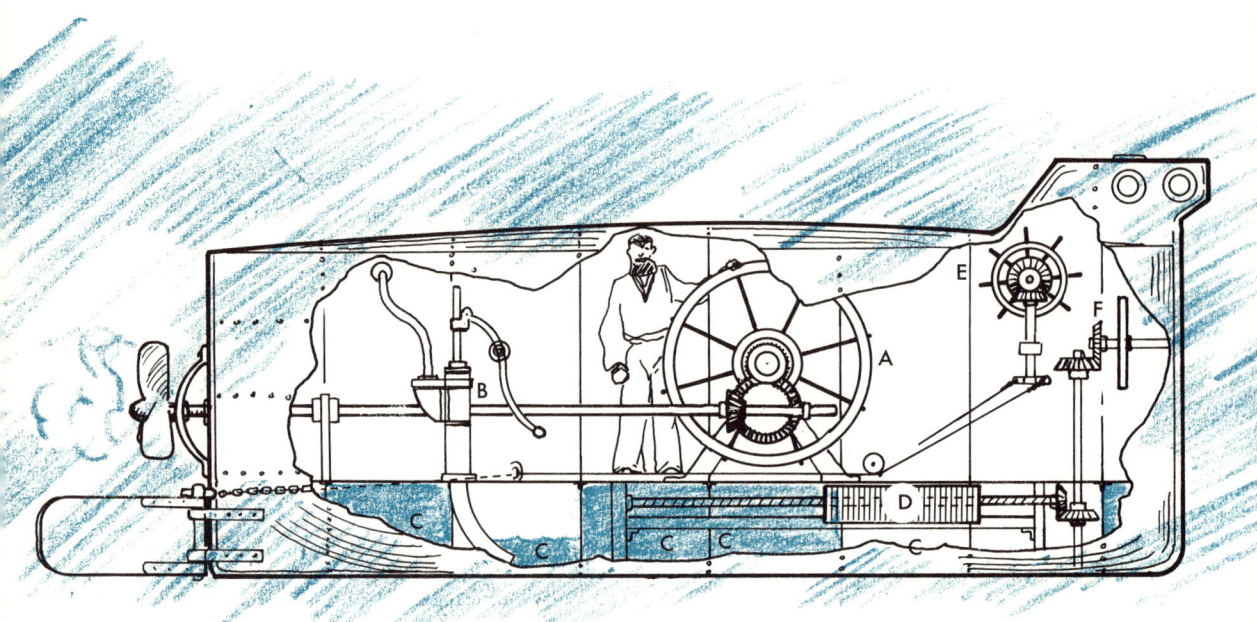

"LE PLONGEUR MARIN"
(Sometimes referred to as "BRANDTAUCHER")

A — Wheel for working propeller
B — Pump for ejecting water ballast
C — Water ballast tanks
D — Weight for altering trim
E — Steering gear
F — Gearing for moving weight, D

world into a blood bath) were at war and a Bavarian named Wilhelm Bauer designed an undersea boat to attack the Danish fleet then blockading the German coast. *Le Plongeur Marin* (*Sea Diver*) as she was called, was a tiny craft—only 26½ feet long and with a beam of 8 feet. She was built of sheet iron, submerged by admitting water to ballast tanks, and was propelled by turning (by hand) a wheel geared to a propeller. She was trimmed—kept in longitudinal balance—and also steered toward the surface, or made to dip beneath it, by moving a weight fore or aft. She carried an electrically fired mine, to be fixed to the hull of the enemy ship.

Like most of the early undersea craft, *Plongeur* could only submerge and maneuver underwater for a very short time and her approach had to be made on the surface. Her appearance, in

December of 1850, scared the blockading ships into keeping a respectful distance but a test dive in Kiel Harbor saw the end of her—and very nearly of Bauer and his two crewmen. Her thin plates gave way and she went down in 60 feet of water. Bauer's method of escape is interesting, as it is the principle used in modern submarine escape hatches. Over the protests of his two-man crew, he admitted more water into the boat until the pressure of the air inside exceeded that of the water outside. The hatches then burst open and Bauer and his seamen shot out and up to the surface. The hull of the craft was discovered in 1887 and was placed in a museum in Berlin.

Criticism of his efforts induced Bauer to offer his ideas first to Austria—then to England. There he worked with a famous engineer, John Scott Russell. But Russell knew little or nothing about underwater vessels and Bauer left in disgust. Russell then designed a boat of his own which proved a total failure. Bauer approached the United States government, was turned down, and finally built a boat for the Russians. *Le Diable Marin* (*Sea Devil*) was 52 feet long and made many experimental dives. One feature of her design was the addition of horizontal rudders. The obstructionism of Russian naval officers finally induced Bauer to give up. He died in 1875—a poor pensioner in his native Bavaria.

Across the Atlantic the American Civil War broke out in 1861 and another blockading fleet, that of the Federal government, was threatening the coast of the Confederacy. Again, attempts were made by the weaker naval power to break the blockade and sink or damage the blockading vessels. The submarines and semisubmersibles devised by the Confederates were known as "Davids"— small boats going out to do battle with the "Goliaths" of the Northern fleet. They were of various types, built of odds and ends of boiler plate (the South was not an industrial power) and about the only thing they had in common was the ingenuity of their design and the bravery of their crews. Some—like the steam "David" in the drawing—were made to submerge only until the top of the superstructure and funnel, which could be telescoped down several feet, were above water. One of these semisubmersibles successfully attacked the Federal ironclad, *New Ironsides*, and damaged her, but the wave of the explosion swamped the boat and only the Captain survived.

Others were cranked by hand and could be made to submerge for short, shallow dives by means of water ballast and horizontal rudders (hydroplanes). Instead of the clockwork or electrically operated mine proposed by Fulton, Bushnell, Bauer, and others, the "Davids" were armed with a spar torpedo. This consisted of a metal canister filled with gunpowder fitted to the end of a long spar. The charge was in some cases designed to be fired by electricity or a firing device operated by a lanyard. Sometimes the canister was fitted with percussion detonators, which exploded—at least in theory—when they struck the enemy's hull. The spar was often rigged so that, as the attacking boat neared its target, the end of the spar could be lowered and the torpedo exploded beneath the water line. The pressure of the water helped confine the explosion to the hull area, thus increasing the damage.

Several successful attacks were made by these ingenious little craft. The most noted exploit was the sinking of the U.S.S. *Housatonic*. The attacking vessel, *Hunley*, was thought at the time to have escaped, but years later divers examining the wreck of the *Housatonic* found the "David" alongside its victim, with the remains of its nine-man crew still aboard. There was no lack of brave men in the Confederate service. This same *Hunley* had already sunk three times and drowned 20 men before she left on her last mission.

While the gallant attacks of the Confederate "Davids" did not affect the course of the war, they did cause considerable concern among the blockading squadrons. Their use also indicated that the day of the close blockade might well be over, and their operations, small as they were, certainly did much to arouse fresh interest in underwater vessels.

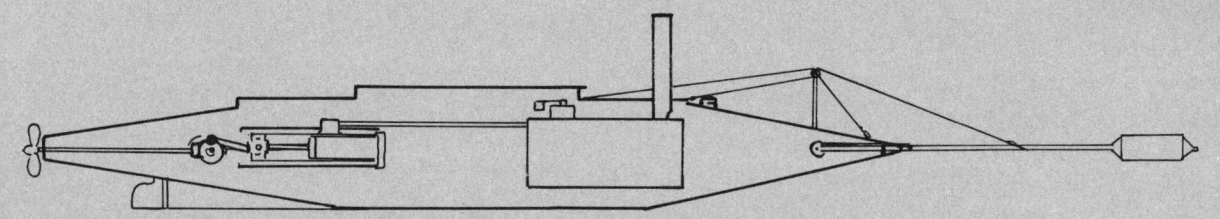

Steam "David" built at Charleston, South Carolina

Length: 54'   Inside diameter: 5' 6"   Crew: 3
Torpedo held about 100 pounds of gunpowder

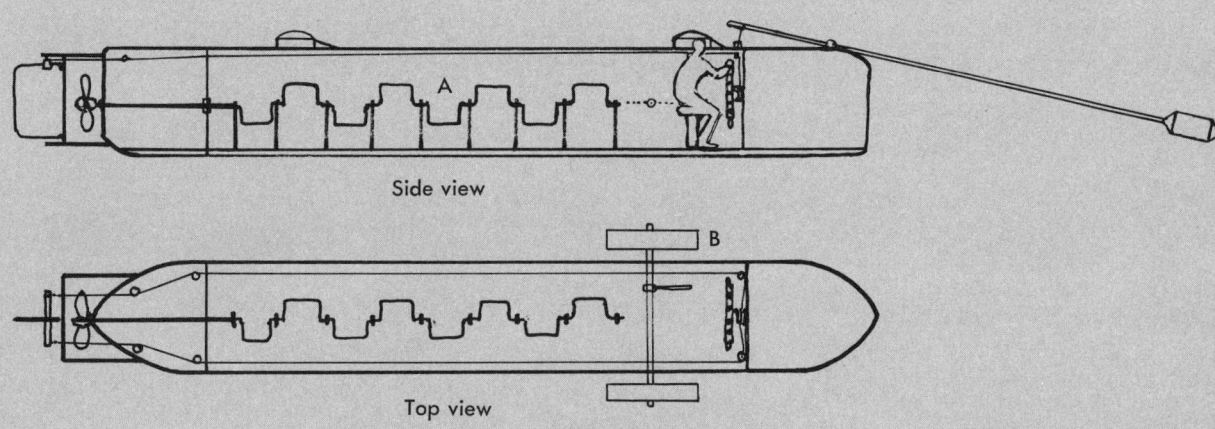

Side view

Top view

Confederate submarine "Hunley" looked something like this. Made partly out of an old boiler, she was about 35 feet long. Eight men sitting at crank, A, turned the propeller. Water admitted to tanks at bow and stern brought her down until only the conning tower hatch was out of water. She could be made to submerge briefly by diving planes, B, operated by a lever.

A spar torpedo of the 1870's. It could be exploded by the trigger device, A, or by electricity.

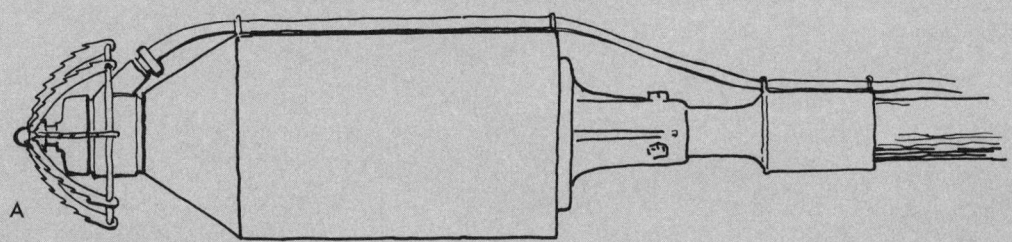

At the same time, the general public was made more receptive to the notion of travel under the sea by such science fiction as the celebrated French author Jules Verne's *Twenty Thousand Leagues Under the Sea*, published in 1870.

Verne's imaginary *Nautilus* was modeled on an ambitious French submarine design, *Le Plongeur*, launched in 1863. Designed by Charles-Marie Brun, this 140-foot vessel was driven by compressed air, stored in large reservoirs. Trim and the angle of descent and ascent were regulated by the admission of water to two cylinders, as well as by horizontal rudders. The vessel was armed with a spar torpedo on a ramlike bow.

Like many early submarines, *Le Plongeur* had trouble keeping proper depth, alternately dipping down or surfacing. The regulating cylinders were too slow in filling and emptying, and even the addition of a vertical propeller did not help. She ended her days as a water tank—but not before she had added her bit to the history of submarine development. She was the largest submarine

"LE PLONGEUR"—1863

Length: 140'   Maximum beam: 20'   Displacement: 420 tons

A — Conning tower
B — 25-foot lifeboat, detachable from inside hull
C — Hatches
D — Vertical propeller
E — Compressed air tanks
F — Engine (surface speed, 5 knots)
G — Ballast tanks
H — Spar torpedo

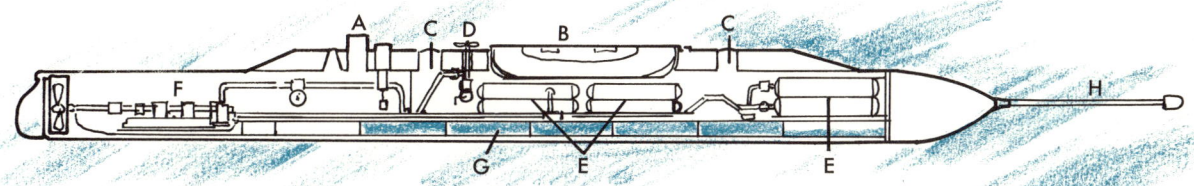

vessel built up to that time, displacing 420 tons, and the first to replace the hand-cranked propeller with one driven by an engine. If she, and others like her, failed to come up to expectations, it was in great part due to the fact that design in many cases was far ahead of technology. Most of the mechanical and engineering achievements we take for granted were either in their infancy or undreamed of. There was no economical way of making steel in quantity; welding, as we know it, was unknown; the practical use of electricity was only dimly understood; the gasoline engine and the diesel were years in the future. So the early submarines, primitive as they seem now, were really miracles of achievement and their designers and builders deserve much credit.

The majority of submarines were designed specifically as war vessels. (Perhaps the fact that as such vessels increased in size and complexity only governments could afford to finance them had something to do with this. And governments in those days were not interested in spending money on undersea vessels for exploration or research.) So it might be a good idea at this point to take a look at the development of the weapons that a submarine might best employ.

First, there is a confusion of terms to be cleared up. During the early and middle years of the last century the word "torpedo" —from the Latin name for the fish known as the electric sting ray,

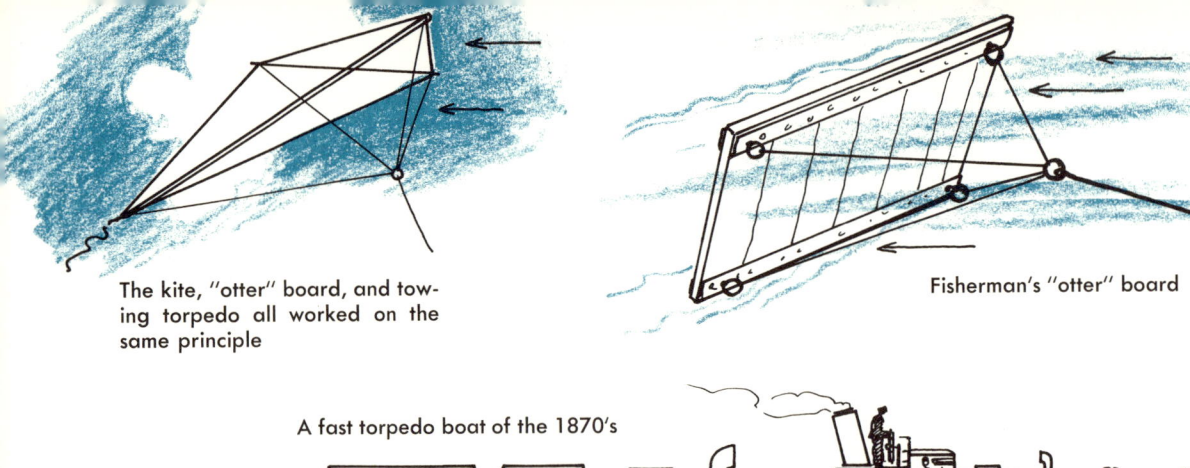

The kite, "otter" board, and towing torpedo all worked on the same principle

Fisherman's "otter" board

A fast torpedo boat of the 1870's

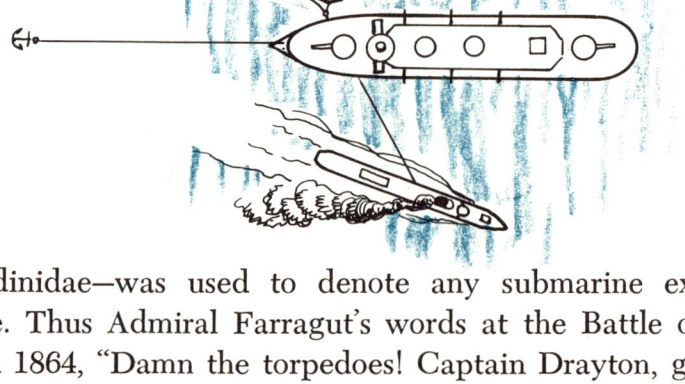

When properly adjusted, the torpedo veered out at some 45° from the towing ship. Attached buoys kept the torpedo at the proper depth. It was fired either by contact or electricity.

family Torpedinidae—was used to denote any submarine explosive charge. Thus Admiral Farragut's words at the Battle of Mobile Bay in 1864, "Damn the torpedoes! Captain Drayton, go ahead . . ." referred to containers of explosive charges anchored below water level—devices which today we call naval mines. The only mobile torpedoes were those charges carried on the end of spars. These spar torpedoes were dangerous weapons to use. The enemy vessel had to be approached within a few yards—a feat likely to prove fatal to the attacker—while the explosion of the charge itself was all too apt to swamp or otherwise destroy the attacking boat, whether a surface craft or submersible.

So the spar torpedo became obsolete and was replaced with devices towed by the attacking craft. A flat board, called an "otter" (used today by fishermen to keep the mouths of their trawls open), caused the torpedo to tow at an angle from the attacking boat. But this arrangement called for great skill, and some luck, on the part of the skipper of the attacking craft. Meanwhile, the increased use aboard ship of rapid-fire weapons—Gatlings, Nordenfeldts, and similar guns—made the chances for survival of a small launch towing such a torpedo pretty slim. The successful use of such weapons by the slow, poorly maneuverable submersibles of the 1870's and 1880's was almost an impossibility.

The answer was a self-contained unit, carrying within itself both explosive charge and motive power. The original locomotive torpedo was the idea of an Austrian naval officer, Captain Lupuis. His torpedo, made in 1867, was designed to operate on the surface, driven by clockwork or steam, and steered from the launching point by guide lines attached to the rudder. The Austrian government thought this too clumsy and ineffective—which it was—and Lupuis consulted an English engineer, Robert Whitehead, then working in Fiume on the Adriatic Sea.

Whitehead began by discarding the steering lines, and finally came up with the forerunner of the torpedoes we are familiar with

TORPEDOES—EARLY AND MODERN
(Drawn to scale)

Whitehead 16" torpedo—1870
Length: 14'   Warhead: 67 pounds of gun cotton   Range at 8 knots: 600 yards

18" torpedo—1900
Length: 16.7'   Warhead: 171 pounds   Range at 30 knots about 1,000 yards

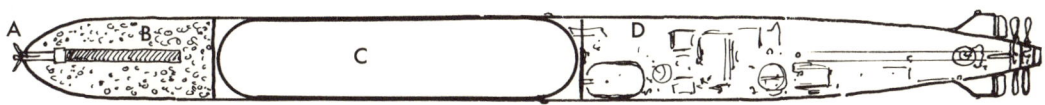

21" torpedo—World War II
Length: 21'   Warhead: 500 pounds plus   Range at 46 knots: 4,500 yards

A — Firing device and exploder. B — Warhead. C — Compressed air chamber. D — Heater, steam turbine, gyro steering engine, depth-keeping device, etc.

today. This 1870 model ran submerged—on compressed air—at a speed of 8 knots, or about 9 miles per hour, had a range of 400 yards, and carried an explosive charge, or warhead, of 15 pounds of dynamite. The Austrian government, for financial reasons, did not purchase the device, but the British did, and Whitehead sold them the manufacturing rights. They continued its development and by 1890 a Royal Navy torpedo could carry a 60-pound warhead at 26 knots for 600 yards.

As the use of the locomotive torpedo spread, fast torpedo boats were designed to launch them. Navies developed quick-firing guns to keep the torpedo boats off, and then torpedo-boat destroyers to sink them before they could attack. So, torpedo boats were built bigger and faster, and the range and power of the torpedo were increased. One development led to a counterdevelopment until it looked like a standoff.

But there was one vessel which could creep up unseen and deliver the torpedo from close range. That was the submarine. A famous British admiral once said that without the Whitehead torpedo there would have been no modern submarines. He may have been right. Certainly the invention of a weapon ideally suited to submarine use spurred the development of undersea boats as nothing else could have. Instead of considering the submarine a crazy craft, designed by crackpots, and manned by madmen, the the more forward-looking naval men began to look upon it as a serious weapon of war—one useful in coast defense and possibly one day even able to venture out to sink an enemy offshore.

STEAM AND STEEL BENEATH THE SEA

The history of submarine development in the last half of the nineteenth century is closely related to technical development in the field of propulsion. There were many promising designs, but the reason that many of them failed was a lack of a suitable engine or engines. Steam was king in those days. Ships, locomotives, machine shops, cranes, pumps—almost everything requiring power was run by steam. It was cheap and it was efficient. But it was also bulky, and because of its need for boilers, furnaces, and smokestacks, it was hardly suitable for an undersea craft. Yet some designers used it, though not with any lasting success.

Hand power had been tried and found wanting, although a few small craft built in the 1870's relied on it. However, the efforts of a few men, sweating over crank handles in the confines of a tiny hull and breathing air which became increasingly foul every minute, could only drive even the smallest hull at very slow speeds for a very short distance.

Le Plongeur's compressed air engines worked—but it was not technically possible to build tanks for holding air under great pressure in her day. Even with her comparatively low pressure, air leaked out, and at best her tanks could only drive her for a short distance. More promising was the use of the electric motor. But at that date such motors were neither particularly powerful or reliable; and they had to be run, at least in a submarine, on batteries—and a lot of them. Several boats were designed to run by electric power, but while it served well underwater—needing no air and giving off neither harmful fumes nor large amounts of heat—a boat could not go very far before her batteries needed recharging. So electricity was not the complete answer.

We accept the gasoline, or internal combustion, engine as a part of our way of life, but in the late seventies it was a newfangled invention and of very primitive design. Although it was relatively compact and fairly well suited to submarine work, in order to exhaust properly, the pressure in the boat had to equal the pressure outside, which precluded any deep diving. However, it was not long before it was realized that the ideal solution was the combination of two motive powers—an electric motor for use while submerged and an internal combustion engine for running on the surface while at the same time spinning a dynamo for recharging the batteries. That was the way the first truly successful submarines were powered, and the way many of them are powered today.

Undersea craft were designed in several countries and the dates of their design, construction, and tests frequently overlap. To avoid confusion and much skipping about from one side of the Atlantic to the other, the major developments prior to the general acceptance of a submarine by the world's navies as a workable weapon of war will be taken up by country. The chart lists them in chronological order.

DEVELOPMENT OF THE SUBMARINE

1850	*Le Plongeur Marin* (Bauer)	Germany
1855	*Le Diable Marin* (Bauer)	Russia
1862-64	Confederate "Davids"	United States
1863	*Le Plongeur* (Brun)	France
1864	*El Ictineo* (Monturiol)	Spain
1872	*Intelligent Whale* (Halstead)	United States
1876	Drzewiecki's Submarine No. 1	Russia
1878	Holland's No. 1 (Holland)	United States
	Garrett's First Submarine	England
1879	*Resurgam* (Garrett)	England
	Drzewiecki's Submarine No. 2	Russia
1881	*Fenian Ram* (Holland)	United States
1884	Tuck's Submarine No. 1	United States
1885	Zalinski Boat (Holland)	United States
	Peacemaker (Tuck)	United States
	Nordenfeldt No. 1 (Garrett & Nordenfeldt)	Sweden
	Goubet's First Submarine	France
1886	*Nautilus* (Ash & Campbell)	England
	Porpoise (Waddington)	England
1887	Nordenfeldt No. 2 & 3, for Turkey	England
	Nordenfeldt No. 4, for Russia	England
1888	*Gymnote* (Zédé)	France
	Peral's Submarine	Spain
1889	Goubet's Second Submarine	France
1890	Two Nordenfeldt-type Submarines	Germany
1892	Pullino's First Submarine	Italy
	Baker's Submarine	United States
1893	*Gustave Zédé* (Romazzotti)	France
1895	*Argonaut Jr.* (Lake)	United States
1896	*Delfino* (Pullino)	Italy
	Morse (Romazzotti)	France
1897	Holland's No. 6 (Holland)	United States
	Plunger (Holland)	United States
	Argonaut (Lake)	United States
1899	*Narval* (Laubeuf)	France

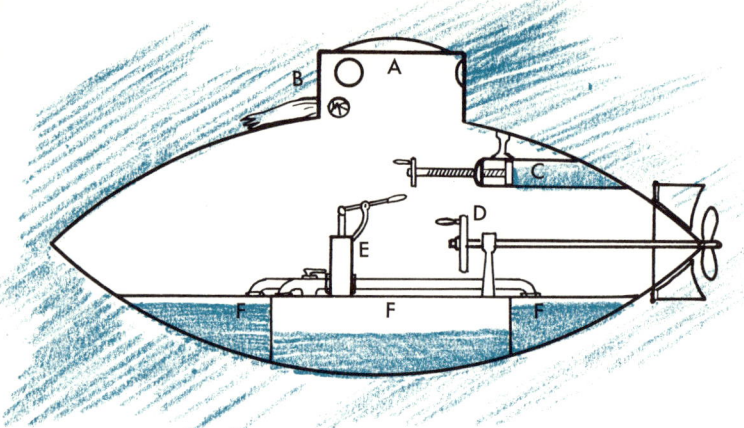

GARRETT'S FIRST SUBMARINE—1878

Length: 14′   Maximum diameter: 5′
A — Conning tower
B — Arm holes fitted with leather gloves
C — Compensating cylinder
D — Hand wheel for rotating propeller
E — Pump
F — Ballast tanks

In England, interest was aroused by the testing of a small submarine built by a clergyman, the Reverend George Garrett, in 1878. This little egg-shaped, hand-propelled craft led to the construction in 1879 of a larger vessel, cylindrical amidships, with pointed ends. This 45-foot *Resurgam*—whose name means "I will arise" in Latin—was steam driven, the first vessel so powered that was designed to submerge completely. When on the surface, the steam unit functioned normally. Before submerging, a head of steam was built up and heat stored in hot-water tanks under pressure. Sufficient heat was stored to provide enough steam to operate for a few miles underwater. When ready to go under, all uptakes, blowers, and furnace doors were, of course, carefully sealed. The boat was moderately successful, although it was lost during deep-water trials.

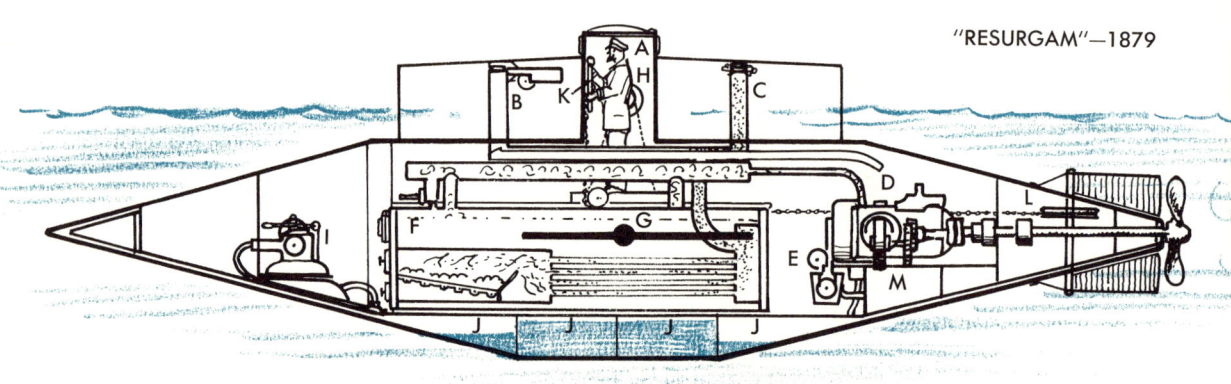

"RESURGAM"—1879

A — Conning tower
B — Air pipe and blower
C — Smoke escape pipe
D — Steam engine
E — Pumps
F — Boiler
G — Horizontal stabilizers (on outside of hull)
H — Wheel for stabilizer control
I — Pump
J — Ballast tanks
K — Steering wheel
L — Rudder chains
M — Hot-water storage tank

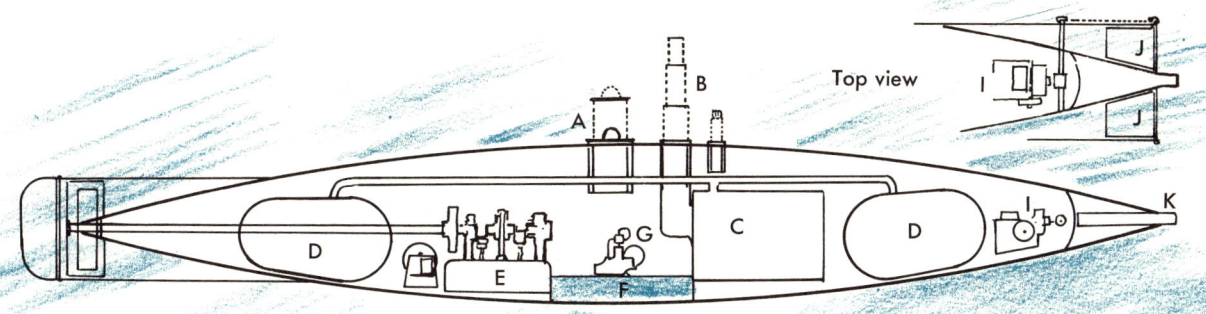

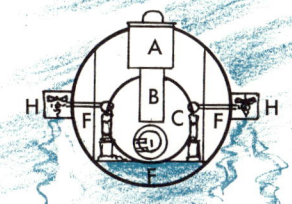

NORDENFELDT NO. 1

Length: 64' Beam: 9' Displacement: 60 tons

A — Retractable conning tower
B — Telescoping funnel and ventilator
C — Boiler
D — Hot-water tanks
E — Engine
F — Ballast tanks
G — Engine for vertical propellers, H
I — Engine for horizontal rudders, J
K — External torpedo tube

Nordenfeldt, a wealthy Swedish gun designer and manufacturer, took up the design, and he and Garrett then built a 64-foot vessel to carry the new Whitehead torpedo. Nordenfeldt favored the same type of steam propulsion used in *Resurgam*, but instead of submerging with the aid of hydroplanes, as had Garrett's boat, two vertical propellers were used. After submerging by admitting water until all but the conning tower was underwater, the boat—which had slight positive buoyancy—was forced down by the screws. This method of submerging—used by several inventors—was never very successful. Nor was driving the vessel under with hydroplanes. Both systems could only be used for shallow dives. Submarines today use the ballast submergence method, coupled with the use of hydroplanes.

Nordenfeldt's boat was tried in Swedish waters in 1885. She ran well as a semisubmersible, with only the conning towers showing, but had difficulty keeping proper depth. One reason for this was the vertical screw arrangement but, in addition, the large hot-water storage tanks and boilers were not sufficiently divided by baffles, so that any inclination of the boat was aggravated by the movement of water. She was sold to the Greek Navy, however,

ASH AND CAMPBELL'S "NAUTILUS"—1885

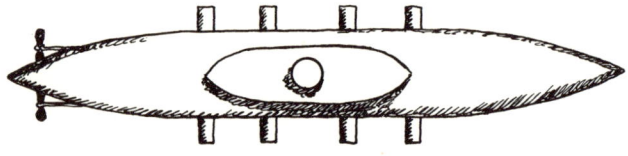

Plan view

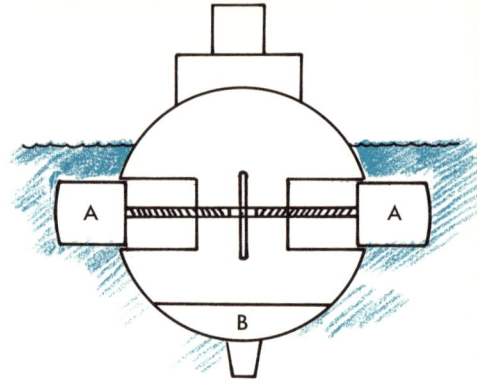

Cross section showing how cylinders, A, A, could be retracted. Filling ballast tanks, B, submerged boat to conning tower level. Final trim was to be accomplished by adjusting the cylinders.

and, no doubt to counterbalance this new weapon, the Turkish Navy promptly ordered two. These were larger, but had the same difficulty keeping depth. The Turks made no use of them. Both were laid up, and soon stripped by robbers of most of their fittings. A fourth boat, 125 feet long, was sold to Russia, but was wrecked while proceeding there under tow. Nordenfeldt's boats were ingenious affairs; his mistake was in persisting in the use of steam for underwater propulsion and vertical screws for submergence.

Several electrically driven submarines were built during the eighties. One, the *Nautilus,* designed by two Englishmen, Ash and Campbell, was the first large electrically powered submarine. She was 60 feet long and her batteries were supposed to supply power for 10 hours at 8 knots. Her method of submergence was original, if not practical. After sufficient water ballast was taken in to submerge the hull except for the superstructure and conning tower, four pistons projecting from each side of the hull were retracted by geared screws. This decreased the displacement—the weight, of course, remained the same—and the boat sank. At the required depth the pistons were worked out until a state of neutral buoyancy was attained.

Fine in theory, the piston-displacement idea never worked well in practice. In one test the submerged submarine stuck in the mud on the bottom and only by moving the crew first forward and then aft could the boat break free from the suction and come to the surface. Unfortunately for the inventors, the Director of Naval Construction was on board and, needless to say, the Navy was not impressed.

A much smaller vessel, the *Porpoise*, designed by a man named Waddington, was tested but found no buyers. And that, for all practical purposes, was the end of submarine development in England until after the beginning of the twentieth century. The rulers of the "Queen's Navee" were no more anxious to see the introduction of a form of warfare which might tend to undermine their great supremacy in surface ships than had Admiral St. Vincent almost a century earlier. It was not until the French began building up a sizeable fleet of submarine vessels that the British naval authorities realized that the submarine was here to stay and that they, too, must enter the undersea race.

The French, at that time on no very friendly terms with the British, were always on the lookout for a weapon which would help counteract the overwhelming might of the Royal Navy. A vessel which could make the close blockade of French ports dangerous, and perhaps impossible, would be very welcome. Any successful development of an undersea boat was sure to receive encouragement by the French authorities.

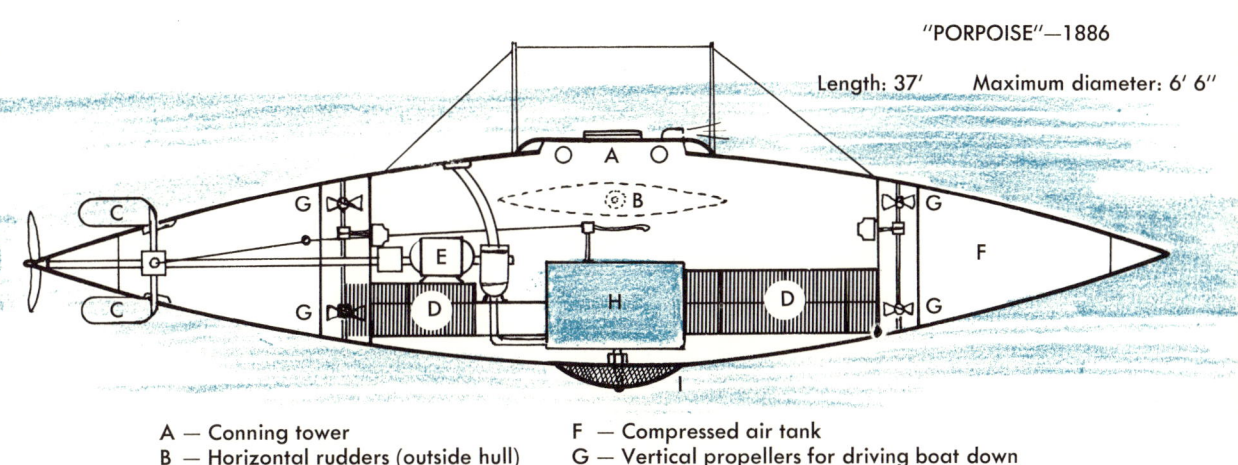

"PORPOISE"—1886
Length: 37'   Maximum diameter: 6' 6"

A — Conning tower
B — Horizontal rudders (outside hull)
C — Vertical rudders
D — Storage batteries
E — Electric motor
F — Compressed air tank
G — Vertical propellers for driving boat down
H — Water ballast tank
I — Drop keel

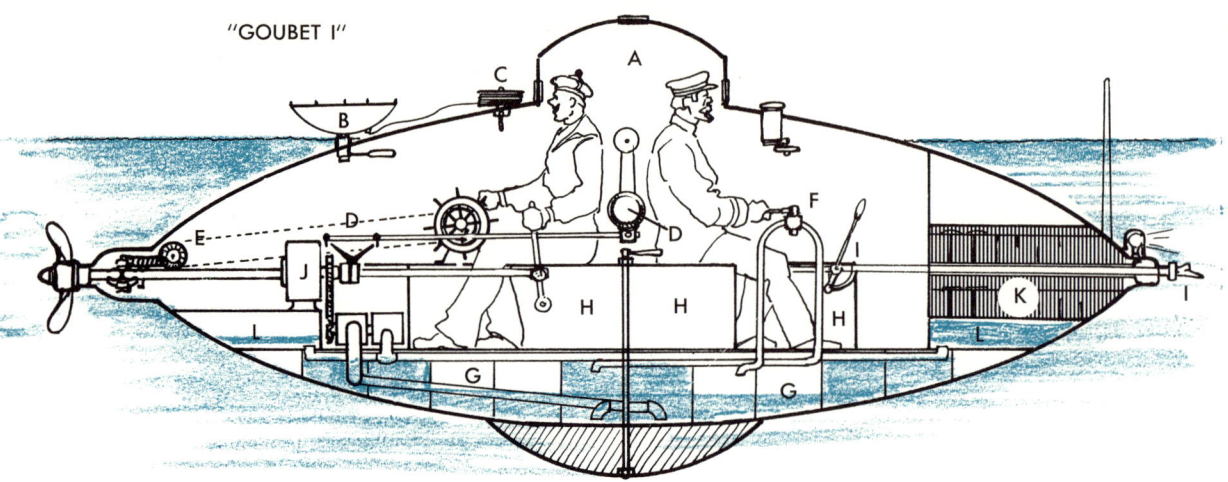

"GOUBET I"

A — Conning tower
B — Mine, to be fixed to enemy hull
C — Reel for wire (to detonate mine electrically)
D — Pendulum device for activating trim tank pumps
E — Gears for moving propeller housing in horizontal plane (for steering)
F — Ballast control valve
G — Ballast tanks
H — Compressed air tanks
I — Net cutting apparatus
J — Electric motor
K — Storage batteries
L — Trim tanks

In 1885, an engineer named Claude Goubet constructed a small submarine, *Goubet I*. The 16½-foot craft was of bronze and carried two men, who sat back to back, their heads in the conning tower. A battery driven electric motor furnished power for a surface speed of 4 to 5 knots and a speed submerged of 3 knots. There was a universal joint in the propeller shaft and the boat was steered by swiveling the propeller. Longitudinal stability was obtained by a pendulum-activated pump which transferred water from ballast tanks fore and aft. A steel rod with a shear attachment could be run out of a tube in the *Goubet's* bow. This was to cut through the nets which warships at that period were hanging out from their sides when at anchor, to protect them from torpedo attack. In Goubet's first boat an electrically fired mine could be released under an enemy vessel. A larger 26-foot version carried two locomotive torpedoes externally.

Numerous tests were made by the French Navy, and both boats performed fairly well. Compressed air tanks allowed the crew to remain submerged for as long as eight hours. But, as in several other experimental craft, the means of attaining longitudinal stability were relatively ineffective, and a submarine which could not maintain her depth was of little practical value. The boat was constantly rising or dipping; on one test she "dipped" to the bottom and was only able to surface by releasing her heavy detachable keel. The accurate discharge of a torpedo was made almost impossible, while the loss of weight when the torpedo was fired further unbalanced the unstable craft. The French Navy decided to wait for something a little better. Goubet died not long after—a disappointed man. But men often profit by the mistakes of others, and Goubet's work was of considerable value to those who came after him.

The Goubet boats were followed by the *Gymnote*, designed by Gustave Zédé. This craft, nearly 60 feet long, was ordered by the French Minister of Marine and was first tested in 1888. Like the Goubet boats, *Gymnote* was powered by an electric motor, the batteries weighing over two tons. Vertical propellers were used to overcome the reserve buoyancy left when the tanks were flooded, and a single hydroplane was used to control depth. This proved insufficient and another was added, but she was still difficult to control, while her batteries gave considerable trouble. Much was learned from her, however, and she was used for many years for experimental and training purposes.

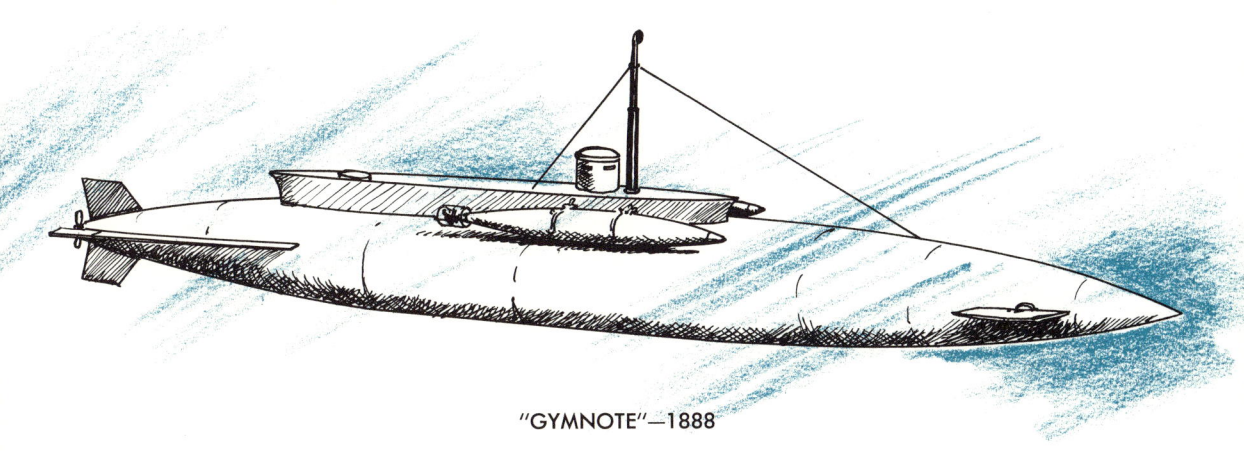

"GYMNOTE"—1888

Showing torpedoes in dropping apparatus

After Zédé died, a new boat, designed by his assistant engineer, Romazzotti, was named for him. The *Gustave Zédé* was a far larger vessel—148 feet long—and her original electric batteries weighed the huge total of 130 tons. She was launched in 1893 and her trials proved her not very satisfactory. The batteries gave much trouble and were finally rebuilt. Also, her depth-keeping ability was poor and extra diving rudders had to be added. By careful experiments, the *Zédé* was finally made into a useful unit of the French fleet.

Meanwhile, a smaller vessel, the *Morse*, was designed. *Morse* was 118 feet long and, like *Zédé*, was driven electrically. Both boats had a bow torpedo tube from which a torpedo was fired by compressed air. In addition, *Morse* carried two more torpedoes in drop mountings on the outside of the hull. The 360-horsepower electric motor drove her at nearly 8 knots, and she had a radius of 100 miles. This was ample for coastal defense work—for which she was intended—and about the limit for an all-electric boat.

Unlike the *Zédé*, which had to rise until the conning tower broke water in order for the commander to see where he was going, *Morse* was equipped with a small periscope. This gave her a fair chance of approaching close to her target without being seen, and in maneuvers *Morse* "torpedoed" several warship targets and was considered a success. Two sets of diving rudders, one forward and the other at the stern, enabled her to keep proper depth without too much trouble.

Realizing that the range of the all-electric boat was hampered by the short life of the batteries, the French had considered a gasoline engine for *Morse*. But the idea was abandoned—perhaps because it was believed that French technology was not capable of building a reliable model of such an engine.

A new boat, *Narval*, was designed by a man named Laubeuf and launched in 1899. This vessel, 111 feet long, was designed with two sets of engines—steam, for surface cruising and recharging the batteries, and electric, for use submerged. With this twin power plant, *Narval* had a surface range of action of 500 miles at 7.5 knots, or 250 miles at 11 knots. She had a double hull, the space between the two being used for ballast tanks. Four diving rudders, or hydroplanes, were used—two forward and two aft. This improved her depth-keeping and gave a smaller angle of glide when she submerged or surfaced.

Narval's considerable range on the surface greatly added to her value as a fighting ship, but she took twenty minutes to dive. It took time to retract the smokestack, cool off unused steam, and secure all draught openings and furnace doors. Also, the heat when submerged made her almost uninhabitable. Partly for this reason, and partly because of the importance of harbor defense in French naval planning at the time, a large number of small all-electric boats were built. The gasoline engine was for some time looked upon with disfavor by French naval designers, and it was not until 1904 that a vessel using a combination of gasoline and electric drives was built.

One of the most advanced submarines, for her time, was designed by a Spaniard, Narcisso Monturiol. *El Ictineo*, as she was named, made many dives and seems to have performed very well.

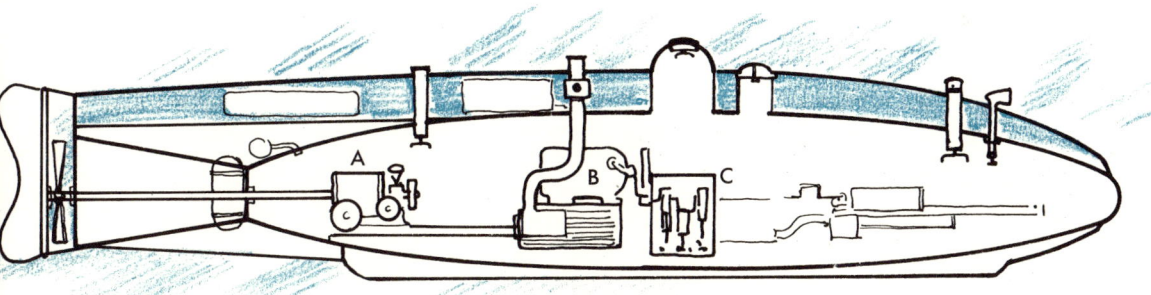

"EL ICTINEO"

Much was claimed of this boat—but little is actually known. The drawing, from which the diagram above was made, shows various pipes, valves, and mysterious gadgets whose use can only be guessed at. Besides her steam engine, A, boiler, B, and pumps, C, water ballast, admitted between her double hull, was expelled by compressed air. She is said to have had an oxygen regenerating plant, a gun, and a device for boring holes in enemy ships.

The next undersea boat built in Spain was launched in 1888, designed by a naval officer named Peral. His submarine was 70 feet long, powered by electricity. It was the type which drove itself down—against a small reserve of buoyancy—by means of vertical screws. As in others of this type, there was difficulty in holding a level when submerged. After creating some enthusiasm, she was turned down by the Spanish naval authorities and soon forgotten. Had the Spanish government been willing to furnish Peral funds for an improved boat, the naval defeats at Manila and Santiago eight years later might not have occurred.

A Russian, M. Drzewiecki, built a small, 14-foot, pedal-driven craft. After water was admitted to lose buoyancy, the fine adjustment to ensure neutral buoyancy and trim was accomplished by a piston and cylinder device. Drzewiecki's boat, completed in 1876, was one of the last to envision the attachment of a mine to an enemy hull—until the idea was revived and used during World War II. To accomplish this the operator, who sat with his head in a glass dome, placed his arms into rubber sleeves and gloves that were attached to the hull.

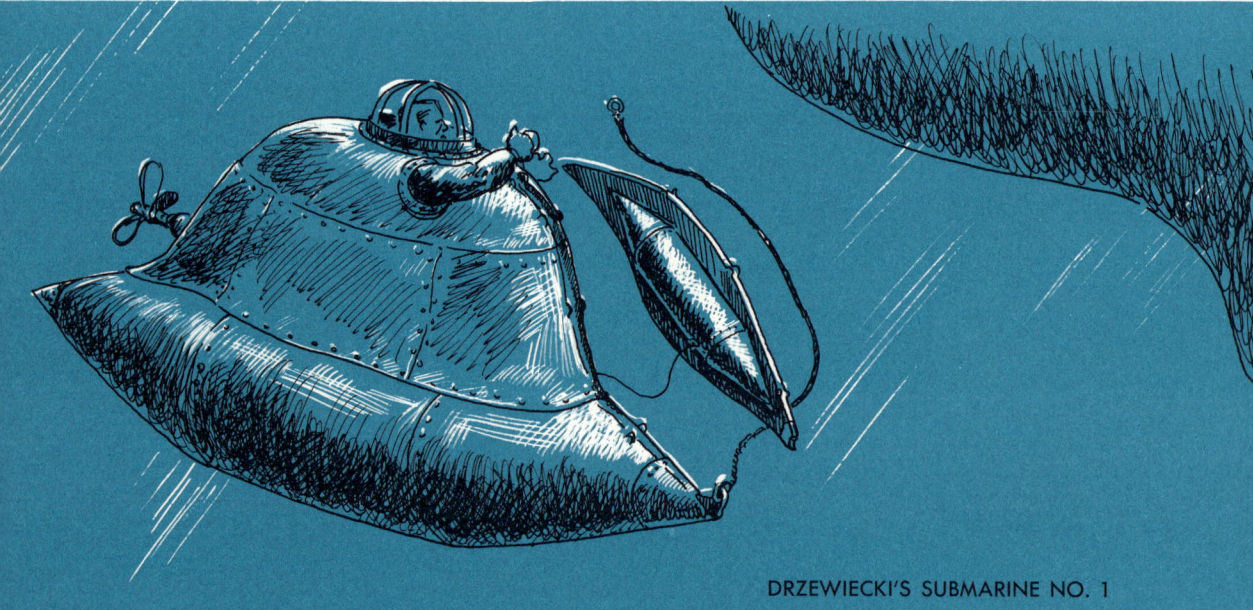

DRZEWIECKI'S SUBMARINE NO. 1

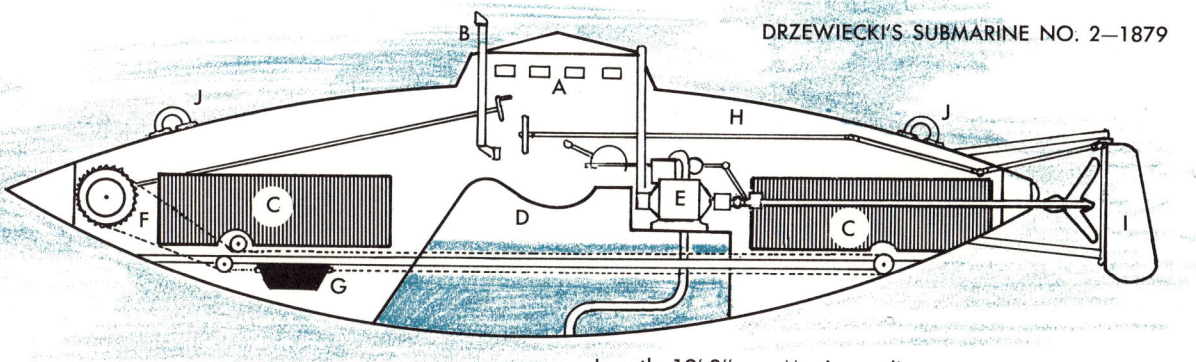

DRZEWIECKI'S SUBMARINE NO. 2—1879

Length: 19' 3"   Maximum diameter: 8' 2"   Electric drive

Originally designed to be driven by foot pedal. Later model electrified. Trimmed by shifting a weight fore or aft. Two mines, not shown, were carried externally. When attached to enemy ship, they were to be fired electrically.

A — Conning tower
B — Periscope
C — Storage batteries
D — Water ballast tanks
E — Electric motor and pumps
F — Chain gearing for moving trimming weight, G
H — Rods for working vertical rudder, I
J — Eye-bolts for lifting boat

A second, and larger, boat was managed by a crew of four, two to work the craft and two to pedal. Two mines were carried, in recesses in the upper part of the hull, to be fired from a distance by electric cable. Fifty of these boats were ordered by the Russian government for coastal defense—the first instance of a large-scale government order for undersea craft. Later, when fairly reliable electric motors and batteries became available, one of these boats was driven electrically. The two pedalers being unnecessary, the crew was reduced to two. The vessels were very slow and were finally abandoned.

In 1901, a more modern vessel, the *Piotr Koschka*, was built from the designs of two Russian naval lieutenants, Kolbasieff and Kuteinikoff. This electrically driven craft was 50 feet long and carried two 18-inch torpedoes externally, in the drop collars patented by Drzewiecki. The *Delfin*, in 1903, was the first of Russian design to be built with a combination gasoline-electric drive.

There was little interest in Germany in submarine boats for many years after Bauer's craft maneuvered off Kiel. In 1890, two boats of the Nordenfeldt type were built. Like the other Nordenfeldt designs, these performed fairly well as semisubmersibles, running with only the upper part of the hull above water. Their diving and underwater capabilities left much to be desired and the type was not repeated. A small electrically driven submarine was built in 1902 which, by that time, had been made obsolete by developments across the Atlantic.

SUBMERGENCE SYSTEMS

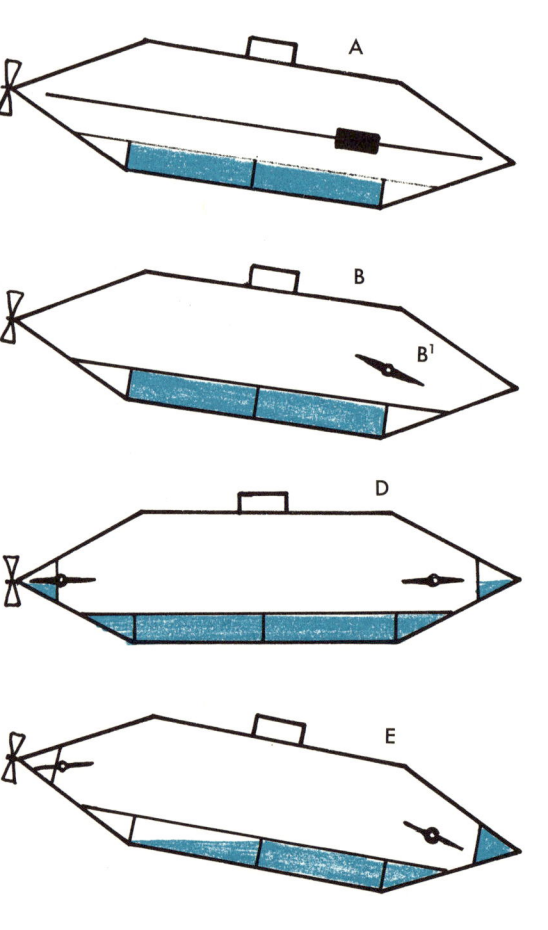

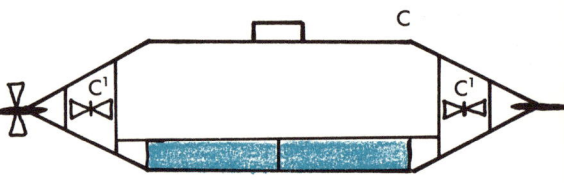

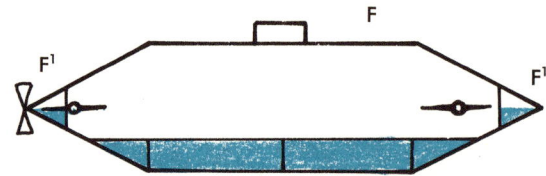

A — Slight positive buoyancy. Boat dived below surface and trimmed by shifting weight. In some, instead of a shifting weight, water was pumped from bow to stern and vice versa.

B — Slight positive buoyancy. Boat dived by use of horizontal diving rudders (hydroplanes, B^1).

C — Slight positive buoyancy. Boat driven under by vertical propellers, C^1. Hydroplanes used for trimming boat (leveling it and making slight adjustments in depth).

D — Negative buoyancy. Enough water admitted to sink boat to desired depth. Then sufficient ballast blown to attain neutral buoyancy (boat neither sinks farther nor rises).

E — Crash dive. Water admitted to forward tanks first. Hydroplanes used as well.

F — Neutral buoyancy. Boat levels by use of hydroplanes and trim tanks, F^1.

THE YANKEES TAKE THE PLUNGE

The exploits of the "Davids" had aroused an interest in such vessels in the United States Navy. A hand-propelled craft, the *Intelligent Whale*, designed by Oliver Halstead, was tested—unsuccessfully—in 1872. But the story of the development of the successful submarine warship is almost the story of one man—John P. Holland.

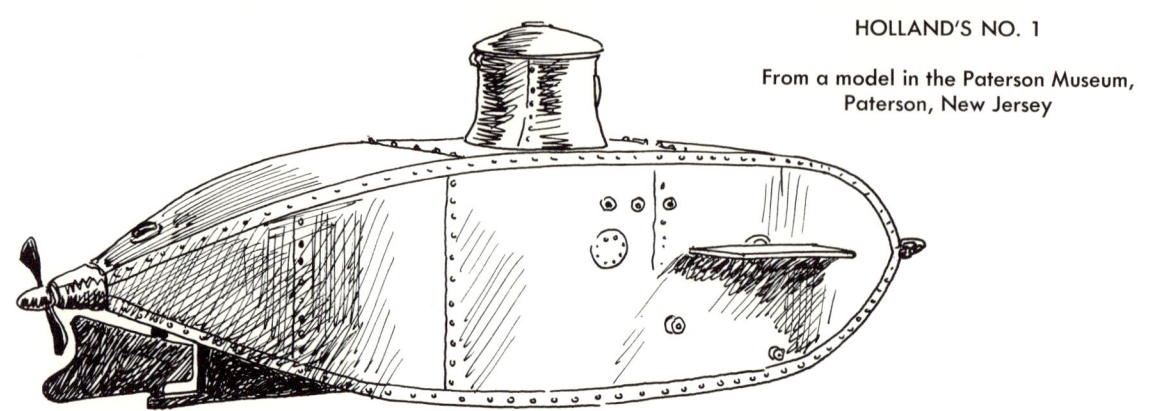

HOLLAND'S NO. 1

From a model in the Paterson Museum, Paterson, New Jersey

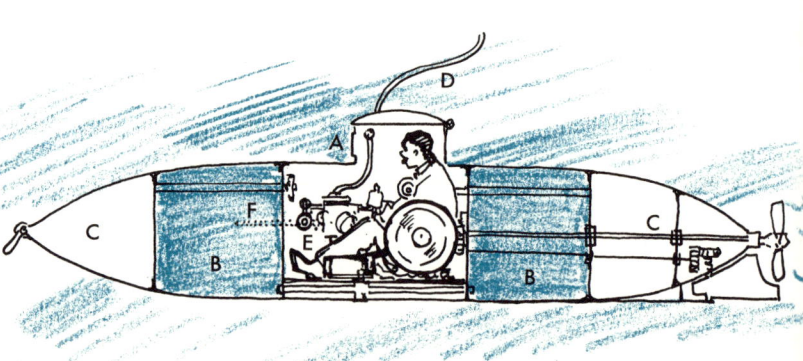

No plans of Holland's first boat exist, but from dimensions and rough sketches, she probably looked much like this.

A — Conning tower
B — Ballast tanks
C — Compressed air
D — Tube leading to boiler of steam
E — Brayton engine
F — Diving planes, worked by hand lever

Holland, an Irish schoolteacher, had little formal scientific education, but he was one of those men who are deeply interested in things mechanical and who often attack a problem with an ability, intelligence, and perseverance which more than makes up for any lack of engineering training. He had become interested in the idea of an undersea warship while still in Ireland, and when he sailed for America in 1873, he brought his sketches with him. In 1875, he submitted his plan for a little one-man submarine to the United States Navy. He received no encouragement from that source, but Irish-American organizations, all violently anti-British, saw in the young Irishman's ideas a means of striking a blow against the hated English.

Financed by contributions from tens of thousands of Irish-Americans, Holland's first experimental boat, the *Holland No. 1*, was built in New York City. The little vessel—probably the smallest ever built—was launched in May, 1878, at Paterson, New Jersey. She was only 14 feet, 6 inches long, 3 feet in beam, and 30 inches high, and had originally been intended to be propelled by pedals. However, one of the new Brayton petroleum engines was installed instead—between the operator's knees, so cramped was the *No. 1*. The engine would not function on its proper fuel, but by running a rubber hose from the boiler of an accompanying steam launch to the cylinder of the Brayton, the boat was able to dive to a depth of some 12 feet and move underwater for a very short distance. Several tests were made; then, Holland having learned all he could from this hull, the motor and fittings were removed and she was scuttled.

Work was begun on a new and larger vessel in 1879 and it was launched in May, 1881. She was 31 feet long and was to carry a crew of three. An improved Brayton engine of about 16 horsepower drove her, and in the bow was a tube for projecting, by compressed air, a shell containing a 100-pound warhead. This missile had an underwater range of about 50 yards and some 300 yards through the air. Ballast tanks, well-designed and well-placed, helped keep the boat in proper trim, and a pair of hydroplanes aft controlled her depth. A newspaperman referred to her as the "Fenian Ram," after the Irish republican society which had helped

"FENIAN RAM"—1883

A — Hatch on top of conning tower
B — Air tanks
C — High pressure air tank
D — Water ballast tanks
E — 15 to 17 horsepower gasoline engine
F — Levers for vertical and horizontal rudders
G — Pneumatic gun

sponsor her. The *Fenian Ram* was a considerable success, but dissension in the ranks of the Irish revolutionary societies ended in one faction "hijacking" her from her berth in Jersey City. She was towed to New Haven, where the harbor master soon forbade her operation, as a menace to navigation. She spent many years in a shed in New Haven and today can be seen in West Side Park in Paterson, New Jersey.

Holland now went to work with George Brayton in an attempt to improve his gasoline engine. At this time he made friends with Lieutenant W. W. Kimball, U.S.N., who persuaded Holland to work for the Navy Department. But Kimball was ordered to sea duty and the Navy Department offered no contract. A Lieutenant Edmund Zalinski of the United States Army, an ordnance expert and inventor of the pneumatic dynamite gun, hired Holland to design a submarine to carry his unique weapon. The "Zalinski boat," as it was called, was not a success. Capital was limited and the boat was finally built of wood on iron frames. Her gasoline engine was inadequate, and her crude periscope arrangement useless. The Nautilus Boat Company, which Zalinski and Holland had formed to finance the boat, was dissolved and the craft dismantled.

There followed some hard times for Holland. In 1888, the United States government announced a competition for the design of a submarine torpedo boat. It must have a surface speed of 15 knots and 8 knots submerged for two hours; withstand pressure to a depth of 150 feet; be easily maneuverable; and have positive buoyancy at all times. Holland's design won, beating Nordenfeldt, and two Americans, Tuck and Baker. But Cramps Shipbuilding Company, which represented Holland, could not guarantee to meet the government specifications and the competition was deferred for the year. The next year Holland won again, but a change of administration led to the funds allocated for the proposed submarine being used for surface craft.

Holland went to work as a draughtsman for a friend who owned a dredging company. Then, in 1892, Cleveland was elected president again and the submarine appropriations were restored. In 1893, in anticipation, Holland and some backers founded the John P. Holland Torpedo Boat Company, but there were numerous delays. Many high-ranking naval officers would have preferred to see money spent on surface ships, and there were interests at work in favor of another—the Baker—boat. It was not until March, 1895, that a government contract was awarded to Holland.

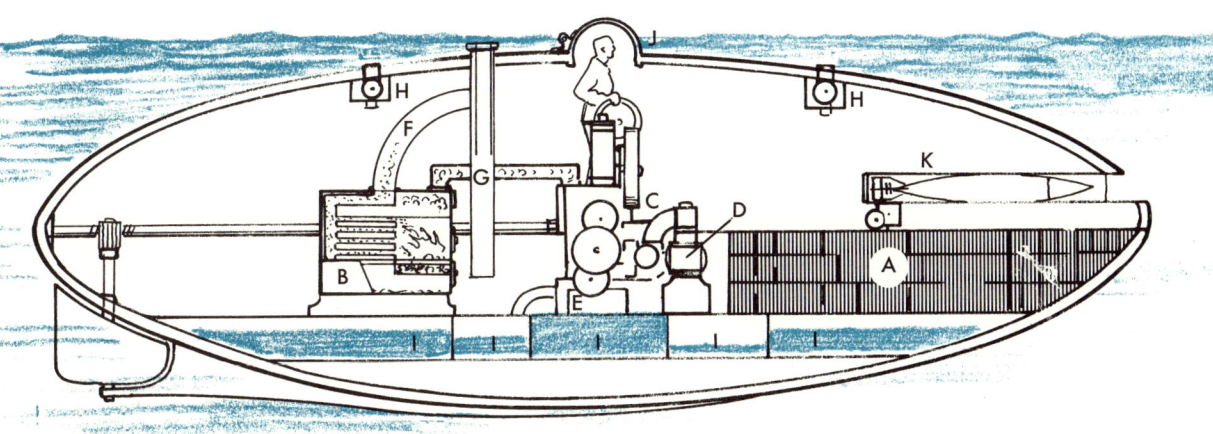

BAKER'S SUBMARINE

The hull, built of 6-inch oak, was 40 feet long and 14 feet in beam. Steam power drove the boat on the surface and charged the batteries for underwater running. Electric fans sucked air into the boat when surfaced. Two propellers, mounted in rotating housings, drove the boat horizontally or vertically. Buoyancy was decreased by admitting water into the ballast tanks. The Baker boat was a failure—like most vessels relying on propellers to drive them down and maintain depth.

A — Storage battery
B — Boiler
C — Main engine
D — Dynamo
E — Pumps
F — Exhaust pipe
G — Telescoping funnel
H — Electric air blowers
I — Water ballast tanks
J — Conning tower
K — Torpedo tube
L — Gearing for rotating propellers
O — Propeller shaft
P — Rotatable propeller housings

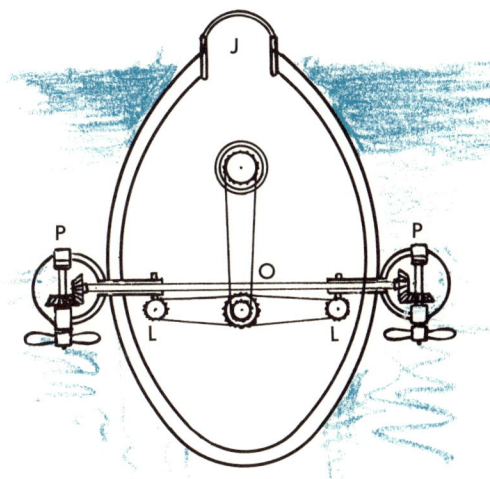

Cross section of Baker boat showing method of rotating propellers in vertical and horizontal positions

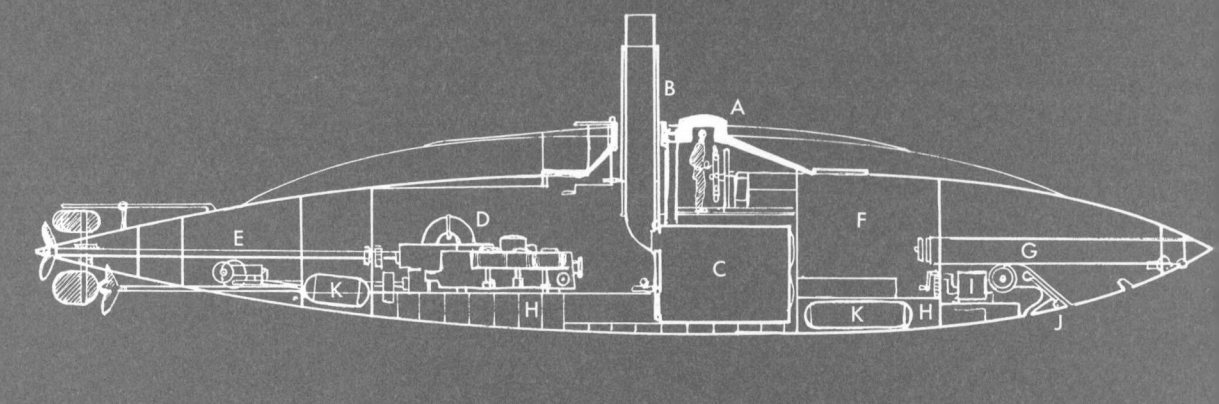

"PLUNGER"—1897

A — Conning tower
B — Retractable funnel
C — Boiler
D — Propulsion machinery
E — Compressor and pumps
F — Torpedo room
G — Torpedo tubes (port and starboard)
H — Ballast tanks
I — Gear for working anchor, J, and opening tubes
K — Compressed air tanks

Holland soon found working for the Navy cramped his style. Modifications were insisted on by those without the knowledge to recommend them, while his own had to be submitted to the Navy Department for approval. The *Plunger*, as she was called, was big —85 feet long and 11½ feet in diameter—and much of her internal space was taken up with the huge steam plant needed to drive her at 15 knots on the surface. A 75-horsepower electrical motor was to give 8 knots underwater. She had triple screws instead of the single screw as originally planned, and two vertical screws for even-keel submergence—which was quite the opposite of Holland's principle of a diving-type boat.

Holland was so convinced that she would be a complete failure that he persuaded his associates to build a new vessel as a private venture and without government interference. Thus, the *Holland VI* came into being.

Fifty-three feet long, with her largest diameter just over 10 feet, she was closer to Holland's ideal "porpoise-shaped hull"— a hull which, he felt, could be driven as fast underwater as on the surface. A 45-horsepower Otto gasoline engine drove her on the surface, and also charged the batteries which powered the 75-horsepower electric motor.

"HOLLAND VI"—launched May 17, 1897

Length: 53.3'   Greatest diameter: 10.3'   Displacement: surface, 63 tons; submerged, 75 tons

45-horsepower Otto gas engine for surface running and charging batteries
75-horsepower electric motor for running submerged
Armament: One tube for Whitehead torpedo; two Holland pneumatic dynamite guns (the after weapon was later removed). Speed: 6 to 10 knots. Crew: Commander, assistant commander, electrician, engineer, gunner, machinist gunner.

A — Storage battery
B — Gas engine
C — Dynamo
D — Main ballast tank
E — After ballast tank
F — Forward trimming tank
G — Conning tower
H — Telescoping air vent
I — Dynamite gun tubes
J — Torpedo tube
K — Oil tank
L — Compressor
M — Dynamite torpedo
N — Aerial torpedo
O — Whitehead torpedo
P — Pumps

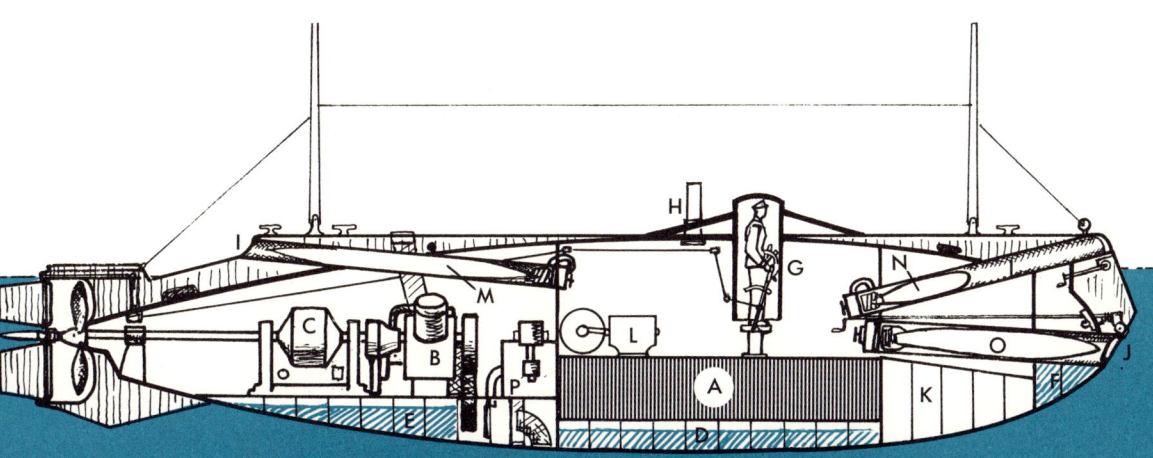

Holland VI aroused a great deal of interest, both at home and abroad. Holland's fame had spread, and representatives of the company had made contacts in several foreign countries. The boat performed well in her tests and an official observer from the United States Navy wrote a favorable report. The Assistant Secretary of the Navy, Teddy Roosevelt, recommended that the Navy purchase the vessel at once—since the country was on the verge of war with Spain—but there was much opposition from antisubmarine naval men. It was not until April, 1900, that the *Holland VI* was bought by the United States government.

Meanwhile, the earlier boat, *Plunger*, had fulfilled Holland's dire predictions. In a ten-minute test run in December of 1898, the temperature in the boiler room rose from 80° to 137° at two-thirds the rated horsepower. It was then planned to convert to internal combustion drive, but the government contract was finally cancelled. Six "improved" Holland boats were ordered, however, from the newly formed Electric Boat Company, with the Holland Company as its major subsidiary.

With United States government contracts and orders from abroad, it would seem that John Holland's position was secure. Actually, he was on his way out. When the Holland Torpedo Boat Company had been formed, he had signed over his plans and patents. Now, with the company's success assured, he was demoted from general manager to chief engineer. Continually pushed further into the background, and forced to watch changes of which he did not approve incorporated into the newer boats, he resigned in 1904. However, Holland was still the big name in submarine boats, and he designed two vessels for the Japanese Navy, larger and faster than the five they had purchased from his original company. He formed his own company to build really high-speed, deep-water craft but he found himself involved in endless patent disputes with the now powerful Electric Boat Company. He became ill and finally gave up and retired. He died, almost forgotten by the public, in 1914.

SIMON LAKE'S "ARGONAUT JR."—from an old photograph

Holland boats had by that time been sold to Great Britain, Russia, Japan, Holland, and Austria. The success of *Holland VI*, the United States Navy's first submarine, had finally taken the undersea boat out of the realm of fantasy and established it as an essential branch in the world's navies.

Holland had only one major American rival. This was Simon Lake. Twenty-six years younger than Holland, Lake was the typical boy-genius type of inventor. At twenty-eight, he had already held patents on some of the hundreds of inventions he would make during his lifetime in many different fields. He had dreamed of undersea travel since reading Jules Verne's classic and, in 1895, his first undersea craft was launched in the Shrewsbury River in New Jersey. *Argonaut Jr.* was a strange little craft. Built of yellow pine, she was 14 feet long and looked a bit like a flatiron. Lake was always obsessed with the idea of a craft which could crawl along the sea floor, and *Argonaut Jr.* had two large wooden wheels, driven by hand. A smaller wheel aft was for steering.

There was a small hand-cranked propeller and a compressed air tank—from a "bankrupt soda fountain"—used with a plumber's hand pump. It was not an impressive beginning, but the *Argonaut*, launched at Baltimore in 1897, was a considerable advance.

Lake's second boat was of iron, nearly 37 feet long, with large iron wheels. She was powered by a gasoline engine, with air supplied through a hose, buoyed at the surface. An air lock at the bow allowed a diver access to the sea bottom. A conning tower with glass ports gave a certain amount of underwater vision. Lake made numerous descents in her, including one with ten newspaper reporters aboard. He enjoyed chugging along the bottom and on one occasion secretly plotted the position of a Navy mine field at Hampton Roads. The Navy had never regarded Lake's undersea boat as anything but a freak, and his disclosure that he had spotted their precious mines merely angered them, instead of demonstrating the effectiveness of his boat.

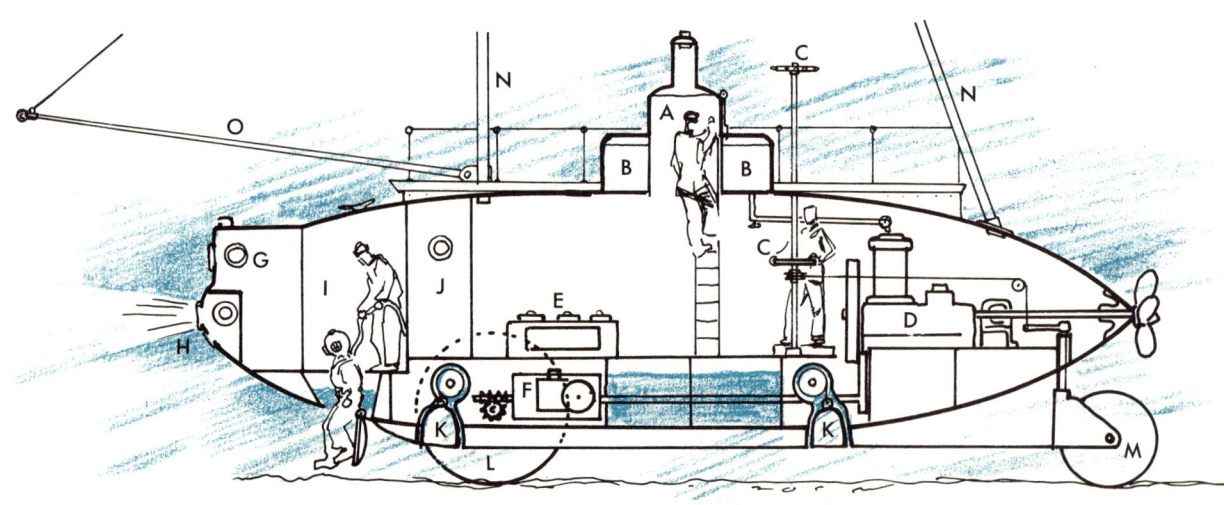

"ARGONAUT"—1897

A — Conning tower
B — Fuel tanks
C — Steering wheels, exterior and interior
D — 30-horsepower gasoline engine
E — Air compressor
F — Pumps
G — Observation compartment
H — Searchlight
I — Pressurized diving compartment
J — Air lock
K — Anchoring weights (could be lowered and boat winched down to bottom or desired depth)
L — Powered wheels
M — Wheel for steering
N — Air intakes and exhaust (hollow spars some 50 feet high, which projected above surface of water)
O — Derrick for salvage work

"PROTECTOR"—launched November 1, 1902

Length: 67' 6"   Beam: 14' 2"   Armament: Three 18" torpedoes; one small rapid-fire gun

A gas engine gave 120 horsepower to each of two propellers. Two electric motors—55 horsepower per shaft—provided power when submerged. Speeds of 10 knots on the surface and 6 to 7 knots submerged were attained. The large steel wheels were raised, lowered, and cushioned hydraulically. A diving chamber in the bow gave access to the bottom. The large conning tower held all controls and the omniscope. A rapid-fire gun was mounted in the small cupola above. A two-piece keel of 10,000 pounds could be dropped with a quarter-turn of a wrench.

A — Omniscope
B — Gun cupola
C — Engine exhaust pipe
D — Hatches
E — Conning tower
F — Torpedo tubes
G — Gasoline engine
H — Dynamo
I — Crew space
J — Horizontal rudders
K — Ballast tanks
L — Storage batteries
M — Drop keels
N — Diving compartment
O — Air lock
P — Anchors
Q — Keel release

In 1902, Lake launched his famous *Protector*. The strides the young inventor had made in seven years—from a laughable-looking wooden box not much bigger than a packing case to one of the most powerful submarines in the world—were tremendous. *Protector* was nearly 68 feet long, with a submerged displacement of 174 tons. (*Holland VI* displaced only 75 tons.) She had twin propellers, three torpedo tubes, a small gun in a hood on top of the armored conning tower, a diving lock forward and, of course, two wheels which could be lowered from the bottom of the hull. Like the Holland boats, a gasoline engine drove her on the surface and electric motors when submerged. A periscope—or omniscope, as Lake called it—was mounted on the conning tower. The air intake of the gasoline engine projected several feet above the deck, so that the boat could run on its main engines with the conning tower partially submerged. A valve arrangement even allowed a very short submergence under gasoline power. She differed from Holland's boats in one major particular. Lake believed in the level-keel method of submergence, while Holland favored the "porpoising" principle, always maintaining a reserve of buoyancy. It is a combination of the two which is used today.

The United States Navy, with several Holland boats built or building, was not interested in *Protector*. However, the Russians and Japanese, at war in 1904, both offered to buy her. Lake sold her to the Russians, followed by seven others. It was not until 1912 that the United States Navy purchased its first vessel from the Lake Torpedo Boat Company.

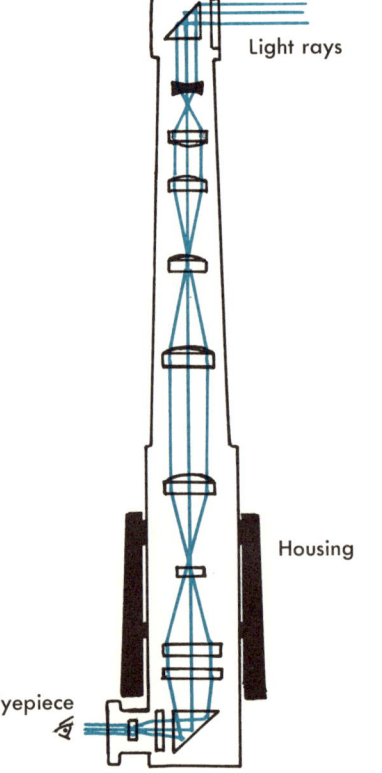

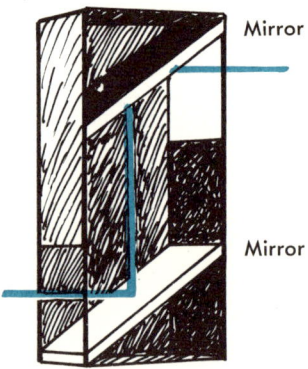

Left: A simplified diagram of a submarine's periscope. Light rays (blue lines) pass through a series of lenses to the observer's eye. The periscope slides up through, and also revolves in, the housing. Watertight glands keep water from entering the hull when submerged.
Above: A box periscope made with two mirrors. This is the periscope in its simplest form.

"SUNK WITHOUT WARNING"

Once the submarine had been accepted as a useful weapon of war, the world's navies set about developing the new vessels at a great rate. Submarines grew in size, speed, and armament, so that by 1914 the little Holland boats of a decade earlier could be considered obsolete. Once the British Royal Navy—which frankly considered its first Hollands as experimental—decided to adopt the submarine, they spared no expense and set about improving the original designs and developing designs of their own. The following table shows the growth of the submarine in the British service up to and during World War I.

Class	Year Launched	Length in Feet	Tonnage, Submerged	Horsepower, Surface	Speed, Surface	Speed, Submerged	Torpedo Tubes	Guns
Original Hollands No. 1-5	1901	63	120	160	8.5 knots	6.5 knots	1	0
A	1902	100	200	600	11.5	7	2	0
B & C	1905	135	300	600	13	8	2	0
D	1908	176	600	1200	14.5	9	3	0*
E	1912	181	800	1600	15	10	5	0*
G	1915	188	950	1600	15.5	10	5	two 3-inch
J	1916	275	1420	3600	19	9	6	one 4-inch
K*	1917	339	2570	10,000	24	9	8	one A.A., two 4-inch

* Guns were added to D and E classes during the war. The K class was steam powered, designed to keep station with the battle fleet.

Other powers also enlarged and improved their submarine fleets. One of the last to do so was Germany—which is strange, because when most people think of submarines, they think of the German U-boats which did so much damage and were, in part,

STEAM-DRIVEN FLEET SUBMARINE OF THE "K" CLASS

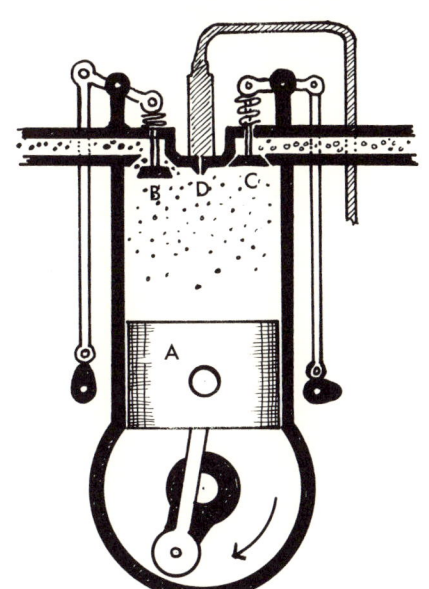

This is the principle on which a diesel engine works. The cylinder (large engines have many) has just fired and the piston, A, is on the upstroke, driving the burned gases past the exhaust valve, B, and out of the exhaust pipe. On the downstroke the intake valve, C, will open and air will be drawn in. On the next upstroke, this air will be compressed and at the top of the stroke a little fuel will be injected through the fuel injector nozzle, D. This will be ignited by the heated air and the cycle will repeat.

responsible for the entry of the United States into World War I. But the heads of the Imperial German Navy were not at first sure that an undersea fleet would serve any useful purpose. Two boats of the Nordenfeldt type had been built in 1890, but were not satisfactory. An electric boat, built in 1902, was not considered a success, either.

Since the failures of early submarines were due in large part to inadequate or unsuitable propelling machinery, the Germans waited until the heavy-oil engine, invented in 1892 by Rudolf Diesel, had been developed to the point where it was reliable. This form of internal combustion engine requires no electric spark to ignite an explosive mixture of gasoline and air. Instead, it relies on the principle that air is heated as it is compressed (the reason that a bicycle pump gets hot as you pump up a tire.) Air drawn into a cylinder on the down stroke of the piston is compressed on the up stroke to about 500-600 pounds per square inch. This heats the compressed air to some 1,000°. At this moment a fine spray of fuel oil (kerosene or something similar) is injected into the top of the cylinder and immediately ignites. The ignited mixture of fuel and air expands and forces the piston down, the burned waste is expelled as the piston rises, and the cycle repeats.

GERMAN U-1—1906

Besides simplicity—no electrical system—the great advantages of the diesel-type engine are fuel economy and the fact that diesel fuel is not highly volatile and explosive as is gasoline. This made for longer cruising range and greater safety.

In 1905, there was considerable excitement in France. A submarine was being built in Germany to plans stolen—so the press said—from France. Actually, the inventor, a M. d'Equevilley, had submitted his plans to the French Navy, who turned them down. Inventors have to live, so M. d'Equevilley showed his designs to the Germans, who bought them.

The new boat so impressed the Germans that, even while she was building, they ordered the first of a great number of submarines which flew the German flag in two World Wars. U-1—U for *Unterseeboot*—was launched in 1906. She was 128 feet long, displaced 236 tons submerged, and had one torpedo tube. Two diesel-type engines turned twin propellers and she had a surface speed of 11 knots. Electric motors gave her a submerged speed of 9 knots. Larger boats followed, and when war broke out in 1914, Germany had some 28 boats of the so-called "overseas" type (as

LARGE OCEAN-GOING U-BOAT—World War I

Length: 275'   Beam: 25'   Speed: 17 knots on surface; 7 knots submerged   Tubes: 6   Crew: 46

opposed to the small short-range boats for coastal defense). It was not a large force. France had some 67, although many were very small and others becoming obsolete, while England had almost as many. But as it became obvious that the German battle fleet had little chance of wresting control of the sea from the Royal Navy, the Germans embarked on a large submarine-building program and some 370 were built during the war. The first warship to be sunk by a submarine since the little *Hunley* rammed her spar torpedo against the *Housatonic* was H.M.S. *Pathfinder*, torpedoed September 5, 1914, by U-21. A week later, the British E-9 sank the German light cruiser *Hela*.

Any lingering doubts among navy men that submarines were capable of striking heavy blows against warships were dispelled on September 22. On that day three big armored cruisers, H.M.S. *Hogue*, *Cressy*, and *Aboukir*, were steaming in line abreast when they were sighted by U-9. *Aboukir* was hit first, and when the others stopped to pick up survivors, they were also torpedoed and sunk. Within a short time a single small vessel of 300 tons had destroyed three ships displacing 36,000 tons, with a loss of some 1,400 lives.

Both Allied and German submarines sank many warships during the war, but the most spectacular achievements of the undersea forces were the German attempts to cut off the great flow of goods upon which Britain depended to survive.

At first, attacks on merchant shipping were made according to international law. Warning was sounded and the crew given time to take to their boats. But by surfacing and waiting while their victim was abandoned, the submarine gave up the very thing which gave her her advantage, the ability to approach and strike her target unseen. So it was not long before U-boats were sinking their victims without warning, often with heavy loss of life. The

After torpedoing a vessel, U-boats often surfaced to question survivors about cargo and destination. Also, few torpedoes could be carried, and many U-boat captains preferred, if possible, to sink their victims by gunfire. The British fitted a few old freighters, even some sail ships, with hidden guns and sent them to cruise where they might be attacked. When these "mystery ships" (or "Q" ships, as the Royal Navy called them) were attacked, the ship was stopped and a well-trained "panic party" hastily abandoned ship. If all went well, the surfaced U-boat came close to finish the freighter off. Then gongs rang, the White Ensign went fluttering aloft, flaps dropped—revealing her guns—and the U-boat was sent to the bottom. On occasion, the vessel was actually torpedoed, or heavily shelled, before the U-boat came within close range. The fighting crew had to remain hidden, sometimes with the ship on fire and sinking under them, before they got a chance to shoot back.

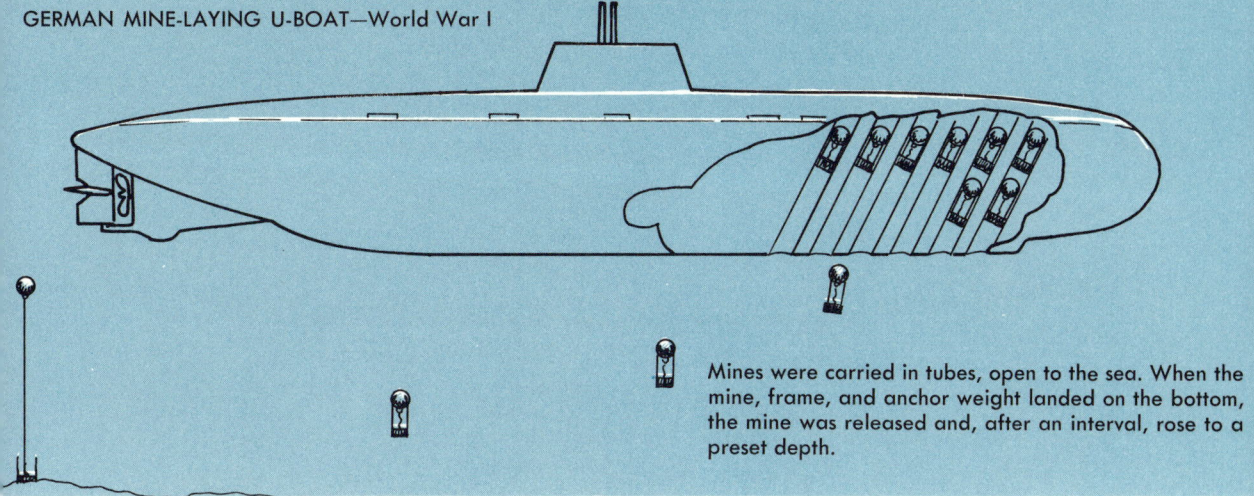

GERMAN MINE-LAYING U-BOAT—World War I

Mines were carried in tubes, open to the sea. When the mine, frame, and anchor weight landed on the bottom, the mine was released and, after an interval, rose to a preset depth.

most publicized sinking, and one which raised a great outcry in the United States, was the torpedoing on May 7, 1915, of the British liner *Lusitania*. She went down with 1,152 passengers and crew, including 114 American citizens. Altogether, German submarines sank nearly 2,300 British merchant ships alone—as well as hundreds of Allied and neutral vessels, plus many warships.

As the war went on, some of the larger submarines were fitted to carry mines. With her warships driven from the surface, there were few areas where German surface mine layers could operate. So to the numbers above must be added the many vessels sunk by mines laid by submarines.

The worst year was 1917. In February and March, sinkings by submarine averaged 23 ships a week. In April, 196 went down. During that period one ship out of four failed to reach port. Ships were finally sailed in groups, guarded by fast warships. These convoys cut down losses considerably and by the middle of 1918 sinkings had gone down, while the number of submarines destroyed had gone up.

During World War I, Germany built some 360 U-boats of various types, some of these very large and capable of cruising across the Atlantic. Nearly 200 U-boats were sunk by British or Allied forces and over 150 surrendered at the Armistice. In four years, the submarine had grown from something of a joke to a deadly menace.

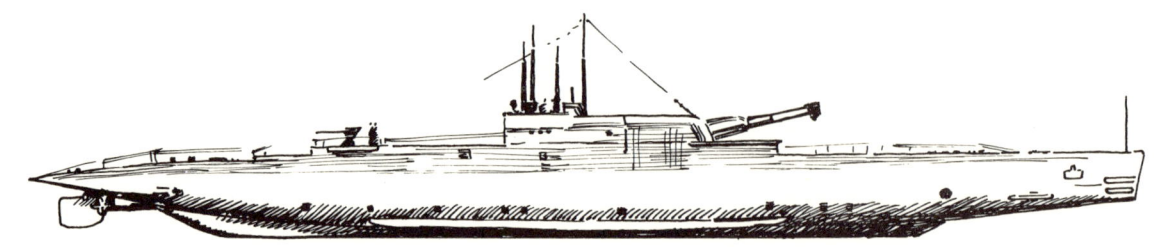

BRITISH SUBMARINE MONITOR M-1

In the period between the two World Wars the basic design of the submarine changed very little. There were some experimental vessels built, but these were exceptions. Among them was the British M-1 class of "monitors." During World War I, submarines had many times popped up to the surface and banged away with their deck armament. Torpedoes were large as well as costly, and only a few could be carried. Also, they were not always accurate beyond a 1,000 yards or so. Why not, the British reasoned, mount a really big gun, one that could do a great deal of damage and be sure of hitting? So the M boats carried a 60-ton, 12-inch gun in a fixed mounting. The gun had a watertight breech, a muzzle cover (or tampion) which could be removed from inside the boat, and could be fired from inside the conning tower. The target was approached in the usual way, the range estimated through the periscope and the gun elevated accordingly. The submarine then blew her main ballast tanks; as the gun emerged, the tampion was opened, the gun fired, and while the 850-pound shell was on its way, the submarine began to dive. From a depth of 30 feet such a boat could surface, fire and be down to 30 feet again in less than 60 seconds. Later, one of these M boats was converted to a seaplane carrier. The gun was removed and a watertight hangar for a small seaplane and a catapult was installed. Another was fitted with a watertight deck casing for carrying mines—boosting the usual number of 16 or 20 carried by the standard mine layer to about 100.

ROYAL NAVY SUBMARINE M-2 LAUNCHING A SEAPLANE

Another British Royal Navy "freak" was the X-1. With a submerged displacement of 3,600 tons, she carried a deck armament of four 5.2-inch guns in two turrets. The French Navy's *Surcouf* was even bigger. She was 361 feet long, with submerged tonnage of 4,300 tons. She carried 10 torpedo tubes, two 8-inch guns in a watertight mount forward of the conning tower, and a seaplane hangar aft. She had a crew of 150 and at 10 knots had a cruising radius of 10,000 miles.

The Japanese Navy built a sizeable submarine fleet, and also produced several which carried a seaplane in a hangar. Many of the Japanese vessels were of the ocean-going type, with a cruising radius of 10,000 miles and were large enough to be used during World War II to carry either landing barges or one or more small submarines on deck. These small subs and underwater "chariots" were developed by several navies.

FRENCH SUBMARINE "SURCOUF"

The small United States submarine fleet had accomplished little in World War I, but a substantial building program in the 1920's and 1930's ensured that the country entered World War II with several classes of modern and efficient vessels. A few were very large and displaced around 4,000 tons submerged. Russia had built up a large undersea fleet but many of them were of the small coastal type. Germany—after first being forbidden to build U-boats at all, and then being limited by agreement, until Hitler declared an end to such obligations—had only a comparatively small submarine fleet in 1939. As we shall see, the undersea boat campaigns of World War II surpassed in scope and savagery those of the "war to end wars" of 1914-1918.

UNITED STATES "R" CLASS SUBMARINE—1918

Length: 186' Beam: 17½' Speed: 13½ knots surface; 10½ knots submerged
Four 18" torpedo tubes; one 3" gun Crew: 29

THE ENEMY IS BELOW

The development of the submarine between 1919 and 1939 was far less spectacular than that between 1899 and the Armistice following World War I. All the same, the undersea vessel of 1939 was deadlier than her World War I counterpart. Although the World War II submarine was little faster, if any, than those of 1917 and 1918, her engines were much more reliable, and she could dive deeper and faster and was better controlled underwater. She could fire larger salvos of torpedoes—which in turn were faster, longer ranged, more accurate, and with a larger explosive charge than in the previous war. Radio communication was greatly improved, as was the apparatus for calculating range and deflection.

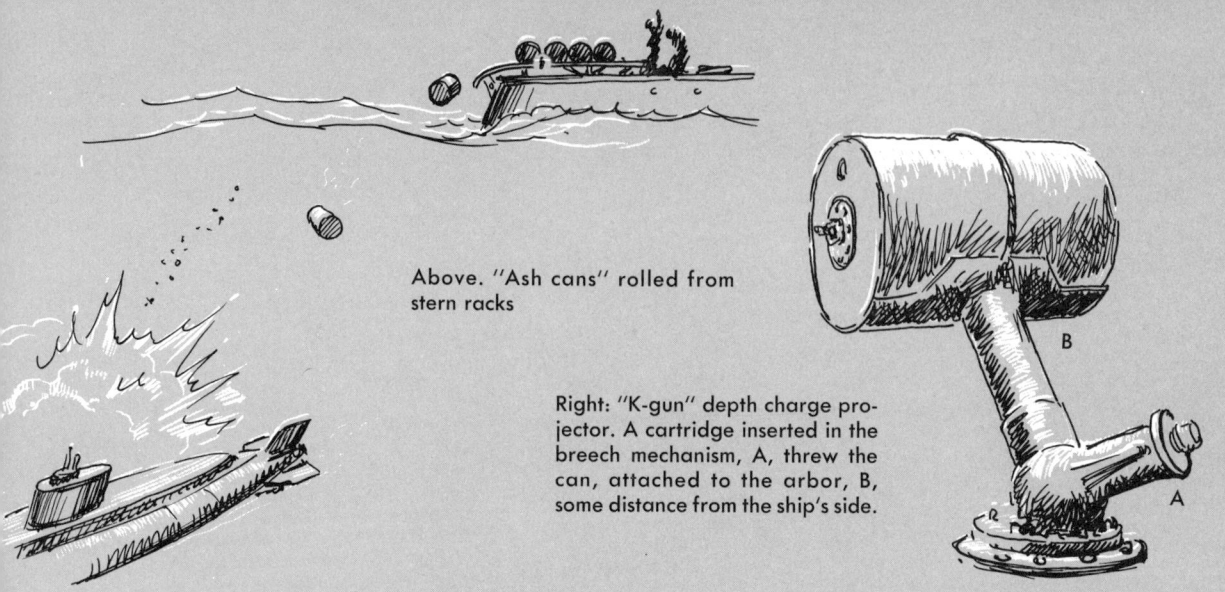

Above. "Ash cans" rolled from stern racks

Right: "K-gun" depth charge projector. A cartridge inserted in the breech mechanism, A, threw the can, attached to the arbor, B, some distance from the ship's side.

As the war progressed, the addition of radar on many submarines greatly increased their efficiency. They could now surface at night and track an unseen target many miles away.

On the other hand, there had been considerable progress in the field of antisubmarine warfare. Depth charges—drums of explosive dropped by a ship and set to go off at a predetermined depth—were invented during World War I by the British. Before that, the only way to destroy a submarine was by gunfire or ramming. Depth charges were now more powerful and ships could carry more of them.

When a firm contact was made on the sonar apparatus, a pattern of charges was dropped, using both the stern racks and the K-guns

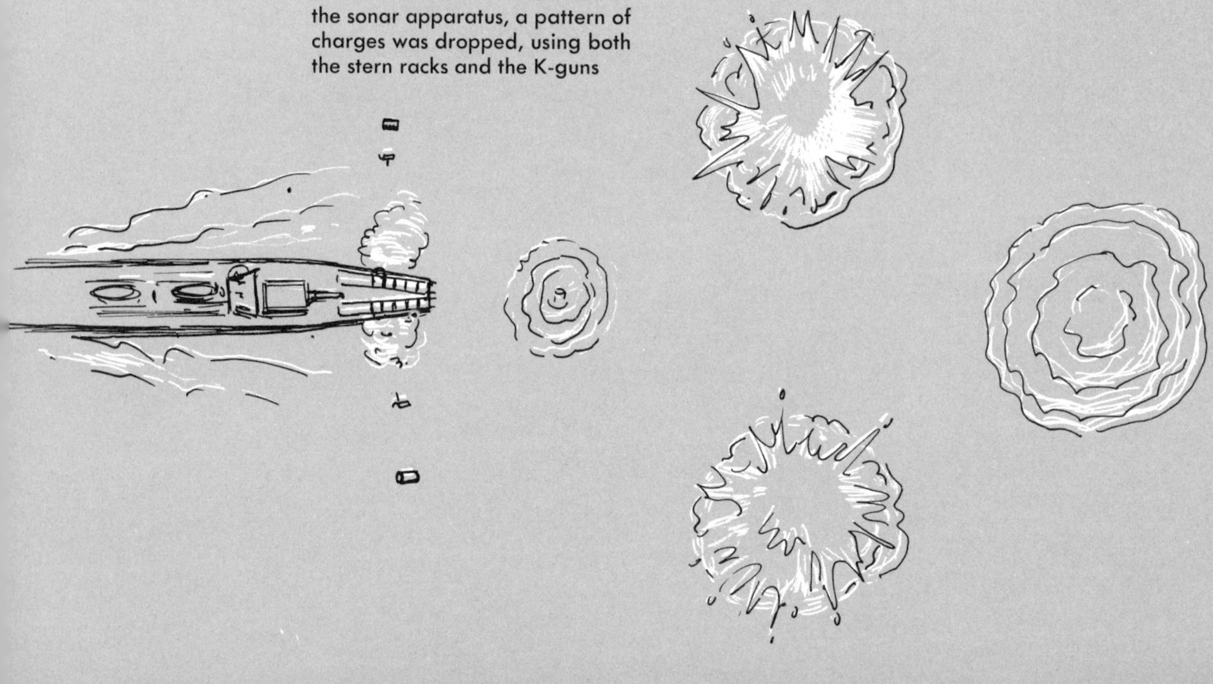

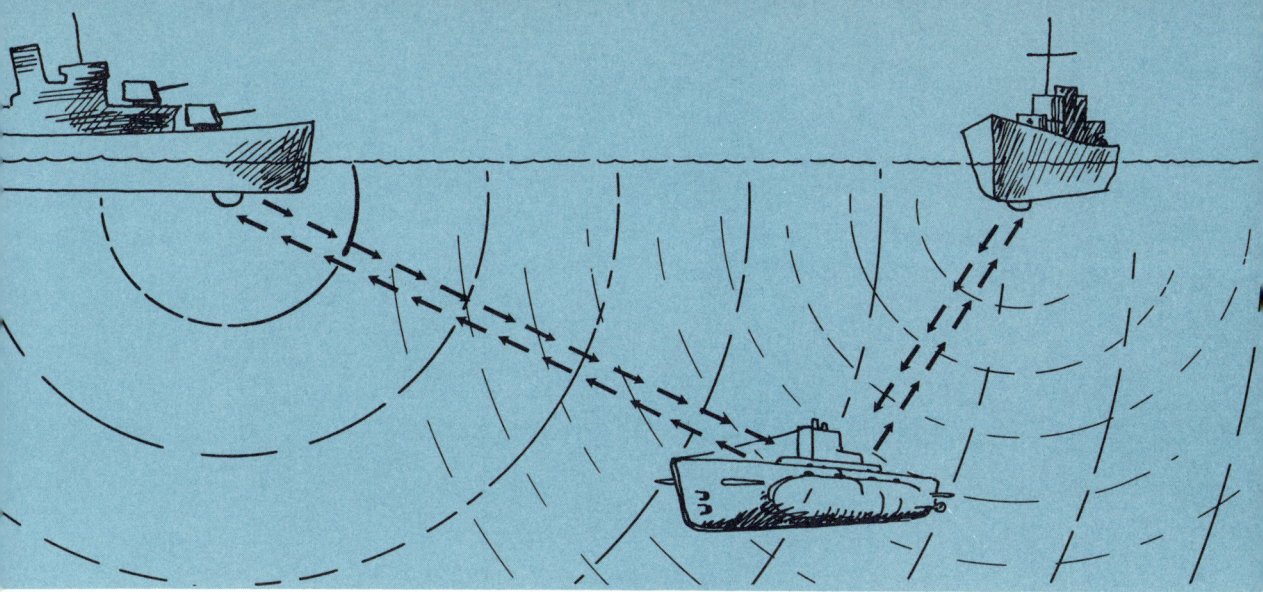

Sound travels underwater, and in World War I acoustic devices known as hydrophones were used to detect the beat of a submarine's propellers. By 1918, scientists had invented an acoustic device which, in effect, "bounced" a supersonic sound wave off a target. The strength of the signal gave the direction, and the time lag between the sending of the signal and the reception of the "echo" gave the distance. The British called it Asdic (from Allied Submarine Detection Investigation Committee), and by 1939 had installed it in many surface ships. In the American Navy it was known as Sonar (Sound Navigation and Ranging). Though the echoes could be returned by schools of fish or dampened by layers of water of different density, the device worked quite well and accounted for many submarine "kills."

Aircraft sank many submarines, and when planes were finally equipped with radar they sank even more. The submarines of World War II suffered from the same disadvantage as those of the previous war—they could only run submerged for a comparatively short time. Then they had to surface—usually done at night—switch on their diesel engines, and recharge their batteries. During that time they were momentarily helpless and, for many, the first warning of an attack was the roar of a diving plane and the whistle of falling bombs. Air attacks are credited with sinking more U-boats during World War II than are surface ships.

TYPICAL FLEET TYPE U. S. SUBMARINE—World War II

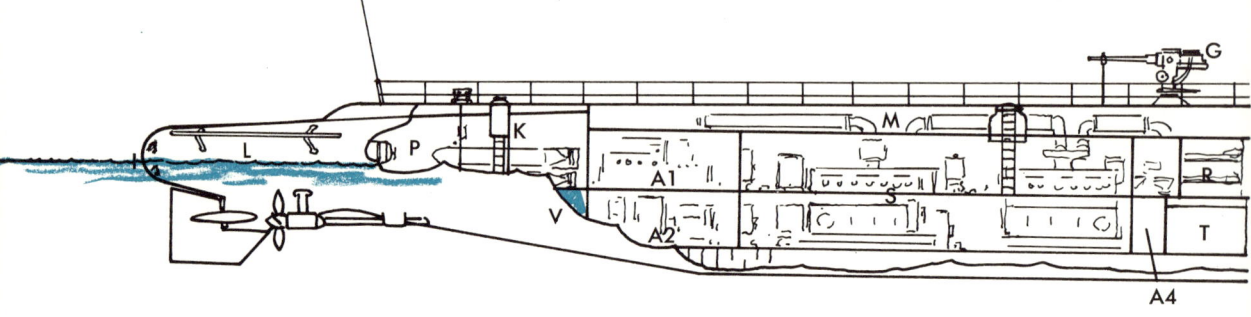

Length: 307'   Beam: 27'   1,500 tons   6,400 h.p.   Surface speed: 20 knots
Ten 21" torpedo tubes; one 3" deck gun; two automatics

The destruction caused by undersea craft during the war was fantastic. U-boats stalked the ship convoys, not singly but in packs often called in by long-range reconnaissance aircraft.

Although Nazi Germany had only 56 submarines when war broke out in 1939, by its end the total had reached 1,150. Of these, no less than 781 were lost, along with some 28,000 men. U-boats destroyed 2,603 merchant ships and claimed 175 Allied naval vessels, most of them British. At one point they came close to cutting Britain's vital transatlantic lifeline. In return, of 632 U-boats lost in action, 500 were sunk by British ships or aircraft.

An even more devastating attack on an island nation's merchant fleet was made by the United States Navy's submarines on Japan. At the time of Pearl Harbor, America had 111 submarines in commission, with 73 building. This rose to a total of 288. Of these, 52 were lost, but a smashing blow had been dealt Japan's economy. That nation began the war with some 6,000,000 tons of merchant shipping and built 3,231,000 more. Of this, United States submarines alone sank close to 5,000,000 tons, as well as sending over 200 naval vessels to the bottom. Of a total of 2,534 merchant ships of over 500 tons sunk, United States submarines are credited with sinking 1,178.

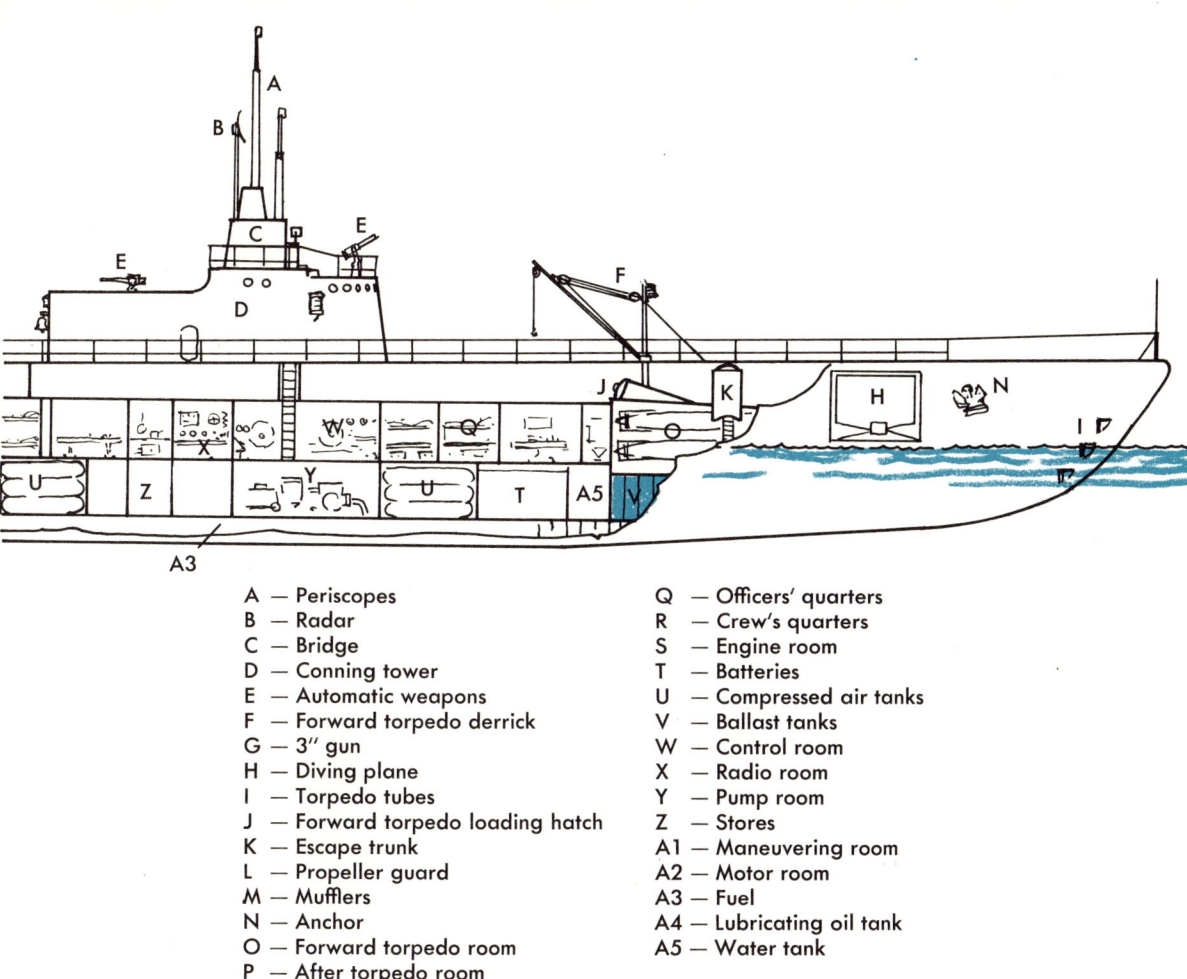

A — Periscopes
B — Radar
C — Bridge
D — Conning tower
E — Automatic weapons
F — Forward torpedo derrick
G — 3" gun
H — Diving plane
I — Torpedo tubes
J — Forward torpedo loading hatch
K — Escape trunk
L — Propeller guard
M — Mufflers
N — Anchor
O — Forward torpedo room
P — After torpedo room
Q — Officers' quarters
R — Crew's quarters
S — Engine room
T — Batteries
U — Compressed air tanks
V — Ballast tanks
W — Control room
X — Radio room
Y — Pump room
Z — Stores
A1 — Maneuvering room
A2 — Motor room
A3 — Fuel
A4 — Lubricating oil tank
A5 — Water tank

The German merchant marine had vanished from the oceans at the outbreak of war, but despite lack of targets, British submarines sank 104 Nazi vessels, as well as many warships. In the Mediterranean they sank over 1,000,000 tons of Italian shipping as well as 42 warships, 21 of them submarines. British submarines also scored many successes in Eastern waters, although this was primarily a United States theater of operations. In all operations, 74 submarines of the Royal Navy were lost.

Japanese submarines were mainly employed against naval, rather than merchant marine, targets. They scored some striking successes, but not enough to make up for the loss of 130 of their submarines, many of them very large.

THE SCHNORKEL

Air for ship's ventilation
Exhaust from engines
Air for engines

A — Automatic closing intake valve
B — Engine exhaust
C — Induction valve
D — Drain, to bilges

Three outstanding developments were made by the Germans in submarine construction during the war. One was the perfection, by a Dutchman named Schnorkel, of an old device that dated back to the early Holland and Lake submarines with the gasoline engines—a tube telescoping up from the deck through which air could be drawn and the diesel engines run while the submarine's hull was submerged. An arrangement of valves shut off the intake if water broke over the top of the tube, as it often did in rough weather. (This was exceedingly uncomfortable for the crew. The U-boat's twin diesels gulped large amounts of air. When the valve momentarily shut, the engines sucked air from the boat's interior, nearly emptying the crew's lungs each time.) The advantage of being able to charge batteries or even proceed under water—except for the top of the schnorkel—driven by the powerful diesels is obvious.

Another "new" development—also going back to Holland's idea of a clean, uncluttered hull that was able to be driven fast underwater—was a hull designed by a German scientist, Helmuth Walter. This hull had originally been planned to carry Walter's revolutionary hydrogen peroxide-fueled engine, then just being put into production. Besides their surface motors, the new boats were given more and improved batteries, which could drive them underwater at 17½ knots for as long as an hour. The normal submerged speed of the average U-boat was only 5 or 6 knots, while the speed of the new XXI type was almost as great as many of the patrol craft guarding the convoys and higher by far than most merchant ships.

Luckily for the Allies, the schnorkel-equipped boats and the new fast XXIs and XXIIIs (a smaller version) made their appearance too late to affect the outcome of the war. The first schnorkels did not get into action until the spring of 1944, while only a few XXIIIs saw service in the last month of the war. Even more fortunately, the Walter hydrogen peroxide-fueled boats—the third major development—remained in the production stage. Had work on their perfection been started earlier, the outcome of the war might have been different. This type of engine gave the submarine, for the first time, a powerful self-contained power plant—capable of operating at high speed beneath the surface without having to rely on short-lived batteries or the schnorkel tube.

The principle behind the Walter engine is the breaking down of hydrogen peroxide—by means of a catalyst, such as permanganate of sodium—into its component parts of oxygen and water. This chemical breaking-down process liberates heat. The water and oxygen then go into a combustion chamber where diesel fuel is injected. The fuel and oxygen mixture burns at a high temperature and generates great pressure. The water is turned into steam and the gas and steam mixture then passes into a turbine which drives the submarine's main propulsion unit.

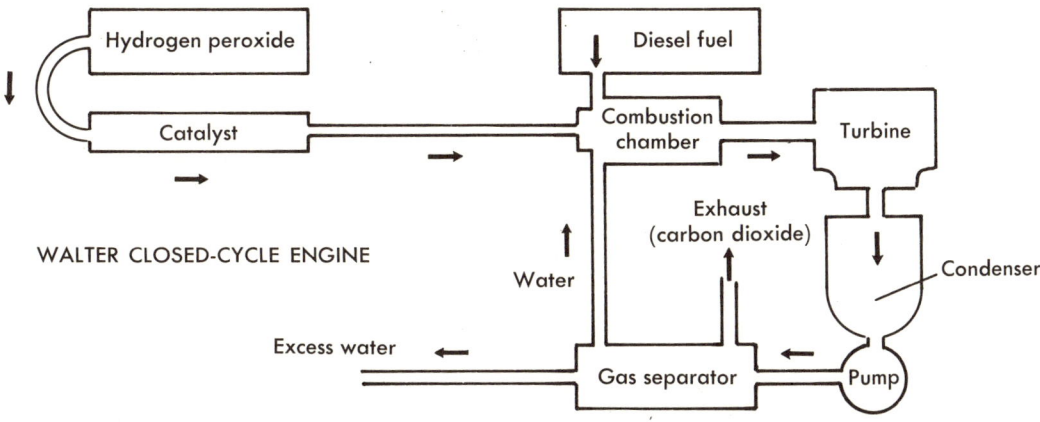

WALTER CLOSED-CYCLE ENGINE

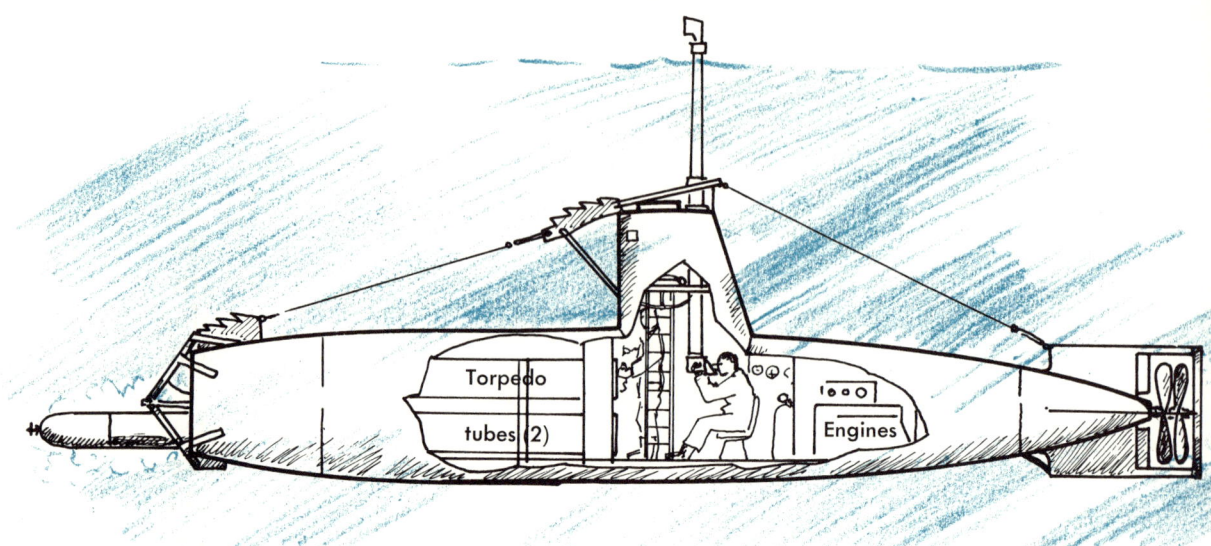

JAPANESE TWO-MAN SUBMARINE—from a model in the Imperial War Museum

In tests, one of the experimental boats reached 25 knots, an incredible underwater speed for those days. In May, 1944, 100 were ordered. They were to carry no deck armament, but were to have ten forward-firing tubes, four in the bow and six amidships. Besides the hydrogen peroxide engine, they were powered with regular diesels for surface travel and electric motors for silent running submerged. Range submerged at full speed was some 160 miles. The first two were to be ready in March of 1945, but Allied bombings had by that time so disrupted production that they never saw service. Such a vessel, if let loose on Allied shipping, would have presented a problem for which the antisubmarine warfare experts would have no answer.

The average ocean-going submarine of 1939-45 was some 300 feet long, with a submerged tonnage of 1,500 tons and a crew of 50. This was a sizeable warship—one not to be risked lightly. Also, because of its size, it was not particularly maneuverable, especially in the restricted waters of bays and harbors. The advantage, under certain circumstances, of using a much smaller boat was obvious. It was less likely to be detected, could be handled easily, and was "expendable."

The Japanese, whose ocean-going submarines were mostly on the large size—some displaced more than 6,000 tons—gave this matter much thought, and shortly before World War II developed the first of their numerous fleet of midget submarines. The Type A boats, some of which took part in the attack on Pearl Harbor, were 78 feet long, 6 feet in diameter, and had a crew of two. They were not "suicide boats"—although their chances may have been slim. They carried two 18-inch torpedo tubes and their electric motors gave them a range of 80 miles at two knots, or a quick 55-minute burst of 19 knots. They were taken within striking distance of their targets on converted seaplane tenders, and could also be carried on the decks of the large submarines. In May, 1942, two submarines launched their midgets at the entrance of Diego Suarez Bay in Madagascar. They sank a tanker, and badly damaged a British battleship. A day later, a similar attack was made in Sydney Harbor, Australia. The target was the U.S.S. *Chicago*, but the torpedo intended for the heavy cruiser sank a small naval auxiliary instead. In both cases, preliminary air reconnaissance was made by a submarine-launched seaplane.

Other slightly larger types followed, with greater operational range, though less speed, and the last model had a crew of five and a diesel for surface running and recharging. They could be prefabricated in two months and 540 were ordered. Intensive bombing slowed production and only 115 were finished by the end of the war.

The Kaiten I Class were true suicide boats. They were actually enlarged naval torpedoes, 48 feet long and 3 feet, 3 inches in diameter. A small periscope gave the sole occupant a chance to guide

JAPANESE KAITEN I SUICIDE SUBMARINE

Later models were larger (54 feet) and faster

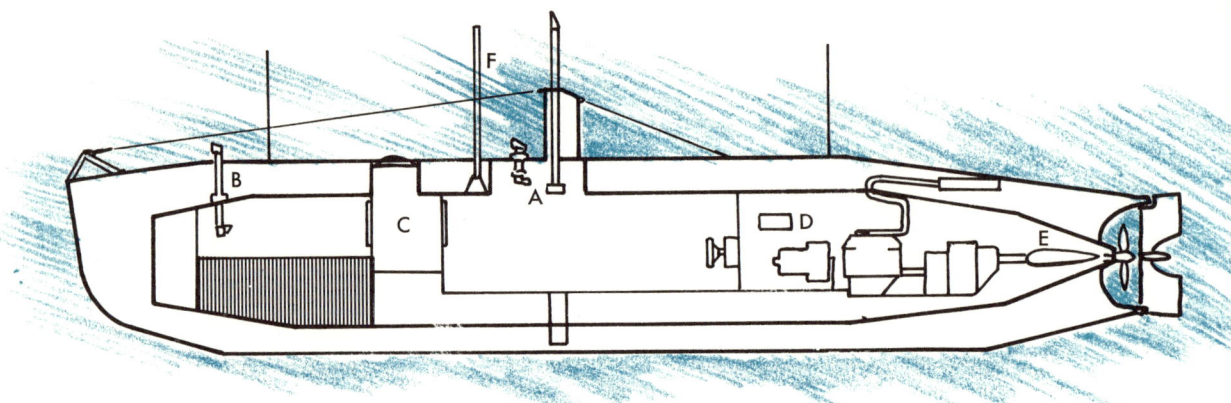

BRITISH X-CRAFT—World War II

A — Periscopes
B — Net periscope (observer lay on stomach over battery space)
C — Pressure chamber and exit hatch for diver
D — Diesel and electric motor space
E — Stabilizer
F — Induction tube (folding) for surface running

his 3,000-pound warhead to the target. These Kaitens were carried, usually four at a time, on the decks of some of the larger submarines, adapted by having part of their deck armament removed and tubes installed from the pressure hull. The Kaiten pilots, their farewells said, could enter their craft from inside the parent submarine and be released, if necessary, some distance below the surface.

The Kaitens proved difficult to handle and the valiant suiciders could count only one success, the sinking of the American fleet tanker *Mississinewa*, in Ulithi Harbor in the Caroline Islands. After that, the vigilance of American antisubmarine forces proved too much for them. Some were seen and destroyed; others were lost when the parent submarines were sunk on the way from Japan to their target areas. In all, the effort cost Japan 96 pilots and eight large submarines.

The British developed their own midgets. Known as X for experimental craft, they were 48 feet long and carried a crew of four. They were to be towed to their starting points behind fleet submarines. There, the regular crew took over from the men who had operated the tow. They carried no torpedo tubes, but had a deck cargo of limpet mines—small containers of explosive which clung by strong magnets to the bottom of an enemy ship—and two 2-ton mines to be placed on the sea bed under the intended victim. The X-craft were designed to penetrate an enemy harbor and attack a vessel at anchor. In September, 1943, several X-craft were towed to the mouth of Altenfiord in Norway, where the mighty battleship *Tirpitz* was lying. One midget went down with her crew on the voyage out, while another had to be abandoned. Of the three boats assigned to attack the *Tirpitz,* one was lost, possibly in a protective mine field. The other two managed to cut their way through the steel antisubmarine nets behind which the battleship lay and drop their charges. The crew of one boat and the two survivors of the other were captured when they were forced to the surface. The mines blew up with force enough to lift the 42,000-ton ship several feet out of the water. She was not sunk, but her turbine engines were shaken out of their beds and she never put to sea again.

In 1945, a similar midget threaded its way into Singapore harbor and blew a huge hole in a Japanese heavy cruiser.

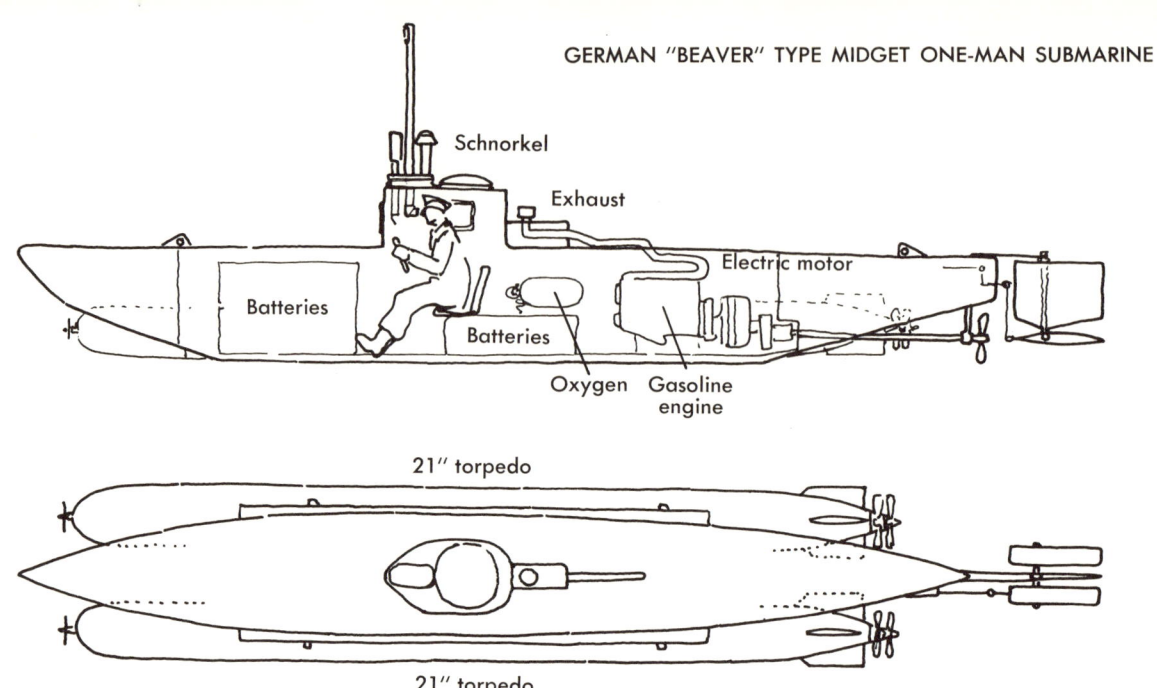

The Germans foresaw an Allied invasion of the continent, and not having the surface fleet necessary to defeat it, designed a number of midget submarines which it was hoped would make such an effort too costly. The first to be operational were single-man boats. They were really two torpedoes, one above the other. The uppermost contained the pilot, who sat with his head in a small Plexiglas dome. When he had maneuvered his craft within range, he released the torpedo underneath, which sped on its way to the target. This type, called "Negroes," could not submerge and could only make some five knots for a comparatively short distance. They were not ready until July, 1944. One sank a British cruiser off the invasion beaches, but once the surprise was over, losses far outran results. Ships fired at anything resembling a canopy; other midgets ran aground or drifted with dead batteries until captured.

A true submarine type, the "Beaver," was brought into service

in late summer of 1944. This was also a one-man boat but was 29 feet long, with a periscope and tiny conning tower and carried two torpedoes outside the hull. They were slow and the pilot, who had only a small compass to guide him, sat with his head barely 36 inches above water level. Despite these handicaps, "Beavers" sank some 90,000 tons of shipping in the River Scheldt, one even torpedoing the lock gates of Antwerp. But losses among the "Beavers" ran to 70 per cent and the later two-man "Seals" also suffered heavy casualties. Altogether, the Germans built some 1,000 midget submarines.

Not all the men who fought under the sea in the Second World War operated inside steel pressure hulls. Wearing some form of oxygen rebreathing apparatus, a number of men on both sides adventured underwater. On at least one occasion, they fought hand to hand beneath the surface. Perhaps because many wore webbed fins on their feet, or because the goggles or faceplates gave them a science-fiction appearance, they became known generally as "frogmen." Some, like the demolition teams—who worked underwater to fasten explosives to the steel and concrete traps planted at low tide on suspected "invasion beaches"—were carried to their work by small rubber boats, often launched from submarines. Other rode "chariots"—weird devices looking much like a cross between a torpedo and a motorcycle.

ROYAL NAVY CHARIOT—1942

The detachable explosive nose of the torpedo was attached by magnets to the hull of the enemy ship and fired by a timing device

Foremost among the "charioteers" were the Italians. Despite many losses, they succeeded in penetrating the British naval anchorages at Alexandria and Gibraltar and sank or disabled several vessels, among them two battleships. Later, British charioteers sank Italian vessels in the same way. The procedure was for the chariot to be carried as near to its destination as possible by submarine, either on deck or in pressureproof hangars. Then the two-man crew, wearing flexible rubber suits and closed-circuit oxygen tank apparatus, climbed into their seats. The chariot's electric motor was started and the machine took off on its mission. All this took place underwater; it was not necessary for the carrier submarine to surface. When the charioteers had reached their target —usually an anchored ship—the nose of the chariot, containing a large charge of explosive with a time fuse, was detached and the crew "dismounted" and fixed it either to the enemy hull or on the bottom underneath it. The divers then mounted up and drove away —if they were lucky.

After the enemy recognized the new form of attack, precautionary depth charging of suspected areas and the periodic inspection of hulls for mines by their own frogmen added to the hazards of the charioteers. It was war of a peculiar kind—fought in the cold and dark, beneath barnacle-encrusted hulls, in seaweed jungles, and rocky harbor bottoms.

"UNDER WAY ON NUCLEAR POWER"

On July 16, 1945, a blinding flash in the New Mexican desert near Alamogordo ushered in the Atomic Age. Men had long dreamed of it—the day when the power of the atom could be unleashed and when a teacupful of matter might be made to release enough energy to drive a mighty liner across the Atlantic. And almost before the reverberations of the bombs which all but obliterated two cities had died away, plans were in progress to harness the atom and make it give up its energy—not in one cataclysmic explosion, but controlled, to produce heat.

Less than ten years later, on January 17, 1955, the U.S.S. *Nautilus* signaled: "Under way on nuclear power." It was fiction turned into fact. Less than a century before, Jules Verne's imaginary *Nautilus*, powered by "illimitable forces ... extracted from inexhaustible sources," had roamed the ocean depths at will, free of any dependence on the world above. Now, another *Nautilus* had also cast loose from the surface and taken to the depths. For the first time there existed a true submarine, capable of making submerged passages of thousands of miles without surfacing.

CUTAWAY VIEW OF ATOMIC SUBMARINE

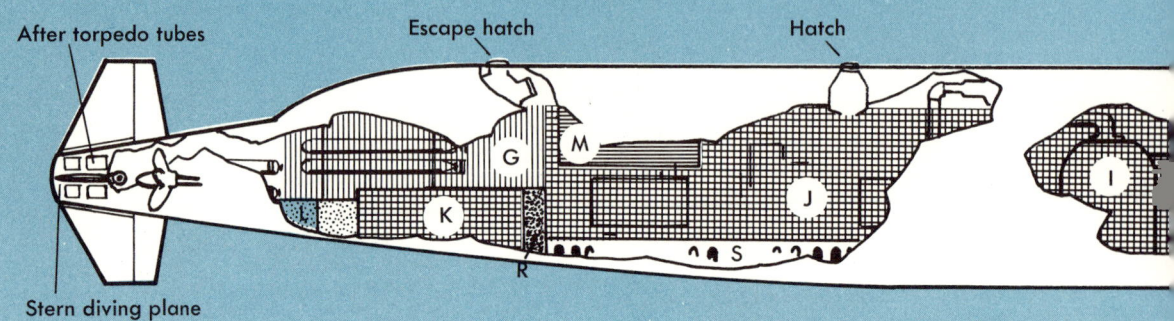

A — Periscope room
B — Radar-sonar and attack center
C — Control room
D — Officers' quarters
E — Crew's mess
F — Crew's quarters and forward torpedo room
G — After crew's quarters and torpedo room
H — Reactor
I — Heat exchanger
J — Engine space, diesel generator, turbines, electric motors, etc.

The actual driving force in a nuclear submarine is the steam turbine. All the nuclear reactor does is supply heat. The rate of fission is controlled by damping rods of some metal which absorb neutrons in large quantities. When the control rods are pushed into the pile, reaction is slowed. When they are withdrawn, greater fission activity results.

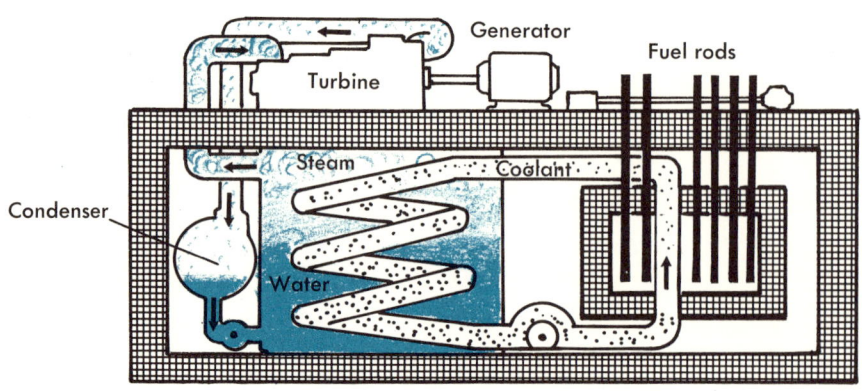

HOW A NUCLEAR REACTOR WORKS

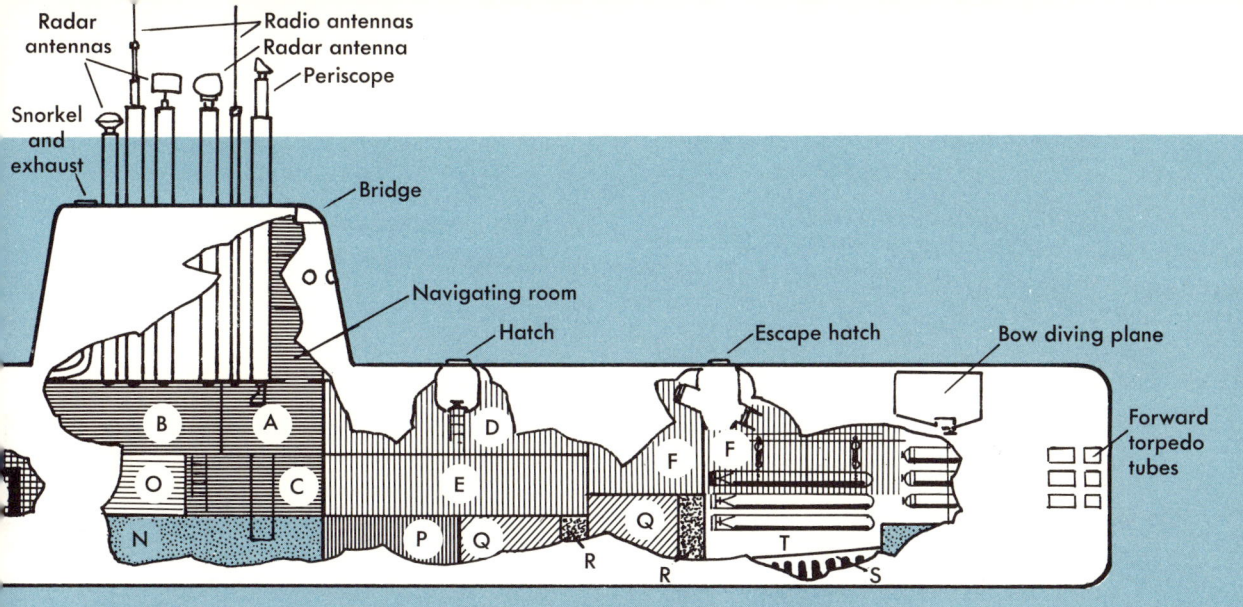

```
K — Hydraulic controls, air conditioners      P — Batteries
L — After trim tank                           Q — Stores
M — Engine control station                    R — Lubricating oil, etc.
N — Ballast tanks                             S — Fuel oil
O — Radar machine room                        T — Spare torpedoes
```

Unlike a regular ship's boiler, the heat of the pile is not used directly to create steam. Instead, the heat is used to raise the temperature of what is known as the primary coolant. This is usually water, but it is kept under high pressure so that it is prevented from flashing into superheated steam when passed through the reactor. This extremely hot primary coolant is then passed through boilers, which produce steam which turns the turbines. The steam, after passing through the turbines, is condensed to water again, and back to the boilers.

Alternate power is provided—diesel-electric or electric from batteries. Oxygen is admitted to the interior atmosphere from pressure tanks, and carbon dioxide and carbon monoxide are held at a low level, being "scrubbed" and filtered in various ways. With the unlimited power available, sea water can now be broken down by electrolysis, supplying oxygen in inexhaustible quantities.

The ability to stay underwater for weeks at a time called for a new system of navigation. There is no point in having a ship cruising secretly beneath the seas if she has to come up to take sights, even if these can now be taken through a periscope. So, along with all the other navigational equipment, the nuclear submarine is furnished with a new and complicated device called SINS (Ships Inertial Navigation System). Basically, in this system, the progress of the ship—every turn or dip or change of speed—is kept track of by instruments, which also absorb all data fed into the system, such as the speed of ocean currents, water temperature, and density. This, coupled with the use of the Fathometer—which bounces a sound wave off the ocean floor, and measures the time of the sound's return and translates it into feet—finds the ship's position with great accuracy.

The new streamlined hulls have now made it possible for submarines to travel faster than many surface vessels. This is because

U. S. FLEET TYPE SUBMARINE—World War II

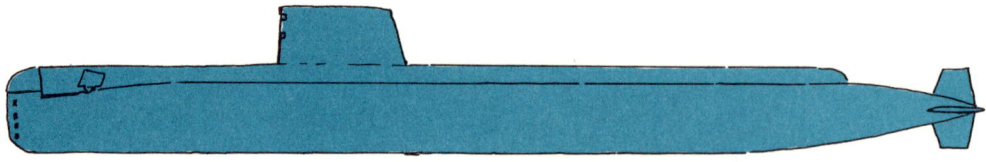

"NAUTILUS"—first atomic-powered submarine

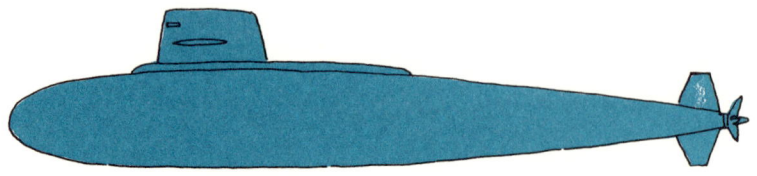

"SKIPJACK"—new streamlined hull plus atomic power

there is no pileup of water such as we see when a surface ship ploughs through the waves. This pileup wastes a great deal of power. A 4,000-ton destroyer's 70,000-horsepower engines can drive her at a maximum of 35 knots. A 4,300-ton submarine needs less than a third of that horsepower to attain speeds of over 35 knots. (How much over is still a secret, but there is talk of 50 knot speeds and better.)

The records made by nuclear submarines are impressive. In one test *Nautilus* ran for more than eleven days submerged, at an average speed of over 21 miles per hour—faster, for over a week and a half, than most previous underwater vessels could go for only an hour. Two years later she was still running on her original fuel load —a little core of uranium.

Nautilus made another famous "first" when in August, 1958, she made a submerged crossing of the Polar icecap, passing *under* the North Pole. Sonar soundings at the spot showed 13,410 feet of water. The 1,839-mile trip under the ice took four days.

Nautilus is only the first of a long line of atomic-fueled submarines. Some, like the *Skate,* are smaller, but designed for underwater speed. Their rounded, porpoise-like hulls would have delighted Holland, and their powerful engines give them underwater speeds of over 35 knots—faster by far than they can go on the surface. Others, like the *Lafayette* and her sister ships, designed to carry 16 ballistic missiles, are twice as large as *Nautilus*—8,250 tons submerged, to her 4,040. In November, 1969, construction began on U.S.S. *Silversides*, the United States Navy's hundredth nuclear submarine.

The Soviet Union has a large and rapidly growing fleet of nuclear submarines—about 60—as well as some 320 conventionally powered vessels. Great Britain has ten "nukes" and France two, with two building.

While at present nuclear submarine development has been confined to vessels designed for war, it is likely that at some future date large submarine cargo ships may be built. There are many advantages. There is more speed for less power, and while a surface ship might be battering her way against mountainous seas and gale-force winds, a cargo submarine could speed quietly along, 300 feet down, unaffected by the storm raging above. Another is the ability to steam under the Arctic Ocean's vast ice fields. This, for instance, not only cuts 4,700 miles off the 11,200-mile voyage from Tokyo to London, but may be of great importance in the development of the newly discovered oil deposits in the far north. There is now a proposal to build a fleet of supersubmarine tankers to transport oil through, or rather under, the icebound Northwest Passage. General Dynamics envisions a 900-foot submarine, with a beam of 140 feet, capable of carrying 170,000 tons of oil!

FREE AS A FISH

Man had come a long way in his conquest of the deep, but the old dream—to free himself from cumbersome helmets and air tubes—was still to be realized. A step toward making the dream a reality was made in 1926, when a French naval officer, Yves Le Prieur, demonstrated a new diving apparatus.

It was really a combination of two devices. One was a helmetless affair with goggles and a noseclip (invented by a man named Fernez), in which the diver breathed air pumped from the surface through a mouthpiece held between the teeth. Le Prieur eliminated the air hose and substituted a compressed air tank strapped on the diver's back. With the Le Prieur-Fernez apparatus, a man could stay down for 10 to 15 minutes.

In 1933, Le Prieur brought out a new model. A mask with a single glass pane—instead of goggles—now covered the nose as well, doing away with the painful noseclip. Air was breathed through the mouthpiece as before, and exhaled into the mask, where it escaped in bubbles. One of the main faults with this piece of equipment was that the air was fed at constant pressure, which was either wasteful, or needed frequent adjustment.

The use of a single-glass mask was a great improvement and was in great part due to the interest in skin diving among those fortunate enough to live on the French Riviera, that sunny spot on the Mediterranean near Nice and Cannes. There, in the warm, clear water and many sheltered coves, the sport of skin diving was born in the early 1920's. One of the prime movers was a successful American author, Guy Gilpatric, whose sea stories appeared almost weekly in the *Saturday Evening Post.* He was one of the pioneers in underwater hunting—with spear and, later, with spring harpoon.

As most people know, trying to see underwater is not only difficult but unpleasant. Light is refracted in such a manner that most vision is lost. Even in ancient days, some pearl divers seem to have used goggles made of tortoise shell polished to near-transparency. Gilpatric puttied up a pair of aviator's goggles so that they fitted closely, and was amazed and delighted with the results. But at any depth, goggles pressed painfully against the eye sockets. And noseclips were uncomfortable. The answer was a tight-fitting, single-pane mask, which covered the eyes and enclosed the nose as well. The mask was now pressurized by the wearer's lungs, and both the uncomfortable pressure on the mask and the equally uncomfortable noseclip were eliminated.

The second step toward the man-into-fish transition was the development of the rubber foot-fin, or flipper. The use of paddles on feet—and sometimes hands, too—is not new. Some natives in the South Seas used crude foot-fins made of woven palm fronds. But it was not until 1935 that the rubber foot-fin was developed by a Frenchman named Corlieu. Flippers not only increase a swimmer's motive power by about 40 per cent, but also leave the hands free.

Step three was the breathing tube—not a new idea either. Ancient peoples used hollow reeds to breath through while hiding from enemies in swamps and rivers. A piece of garden hose, lead-

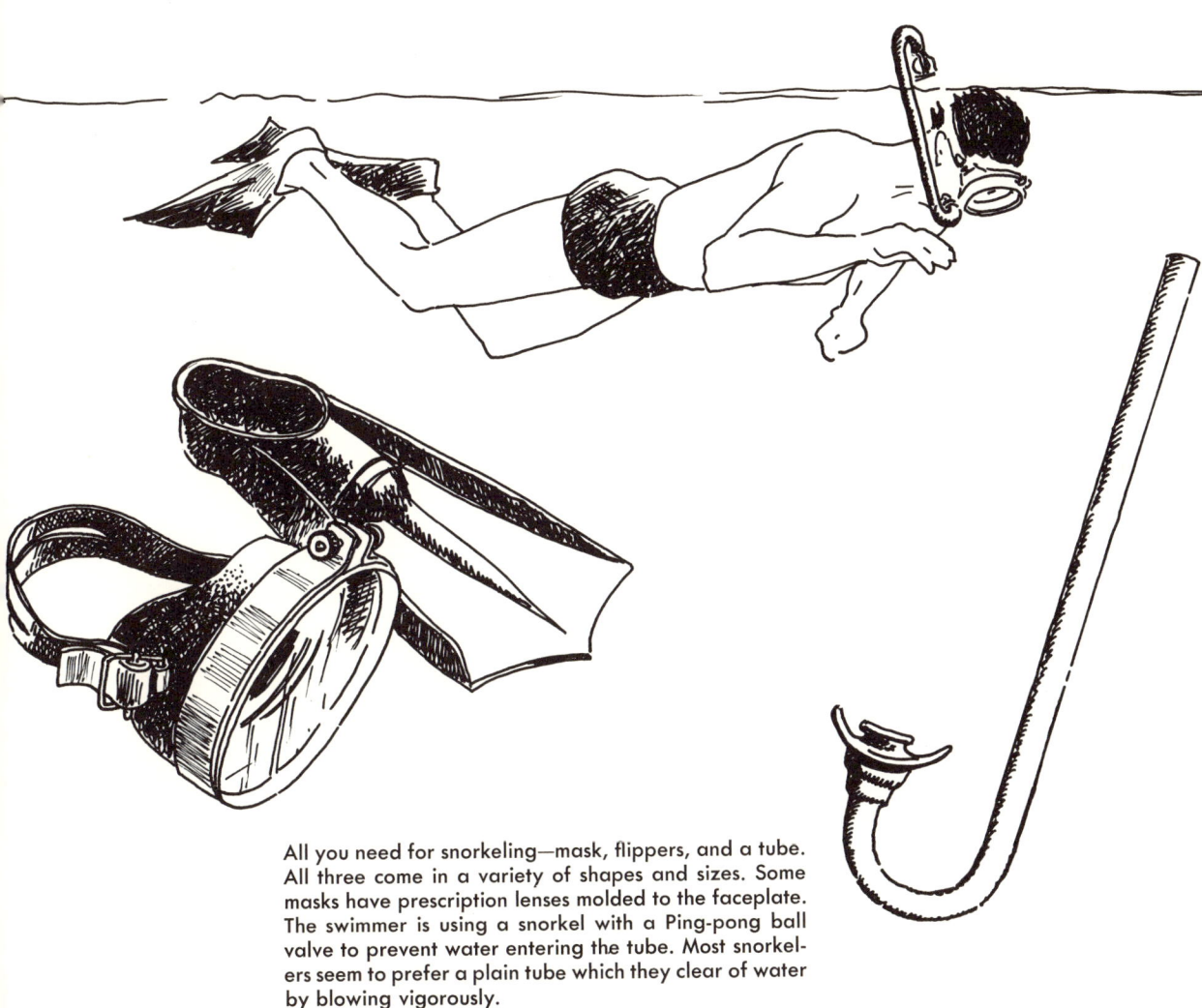

All you need for snorkeling—mask, flippers, and a tube. All three come in a variety of shapes and sizes. Some masks have prescription lenses molded to the faceplate. The swimmer is using a snorkel with a Ping-pong ball valve to prevent water entering the tube. Most snorkelers seem to prefer a plain tube which they clear of water by blowing vigorously.

ing to a mouthpiece, showed what could be done. Now there are dozens of such devices on the market. The "snorkeler" has a choice of many masks and tubes, some with shut-off valves. Nor is the swimmer confined to paddling face down on the surface—although in clear tropical waters this can open up an amazing new world. With a lungful of air and a few beats of his flippers, the snorkeler can go below—limited only by the length of time he can hold his breath. On surfacing, a good blow clears the tube of water—like a whale spouting.

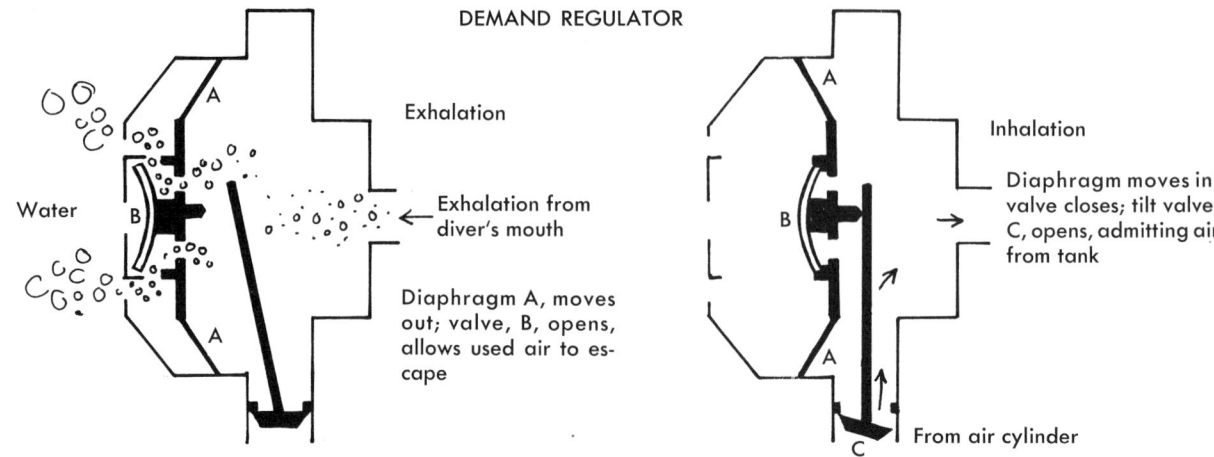

The combination of the face mask, flippers, and Le Prieur's air tank apparatus gave skin divers a foretaste of what was to come. But it was not until another pair of Frenchmen—naval officer Jacques-Yves Cousteau and Emile Gagnan, an engineer—perfected the now famous aqualung that man really began his return to the sea from which he had emerged so many millions of years ago.

The success of the aqualung and others like it lies in the valve which regulates the flow of air from the tanks to the mouthpiece. If the pressure of the air taken into the lungs does not equal the pressure of the water at any given depth, either the diver will not be able to breathe (the water pressure will not allow him to inflate his lungs) or too much air will come through and be vented off and wasted. The aqualung has a "demand" system designed to deliver exactly the amount of air needed to balance water pressure. As the diver descends, more air is released; as he ascends, the flow of air is reduced. Exhalation is through a nonreturn valve, placed as close to the intake as possible so as to be at the same pressure.

The exploits of Captain Cousteau and his fellow divers have been well publicized in books, pictures, movies, and television. The comparative simplicity and low cost of gear of this type have encouraged hundreds of thousands to enjoy the delights of underwater sport and exploration, and the number of enthusiastic scuba (self-contained underwater breathing apparatus) divers is on the increase.

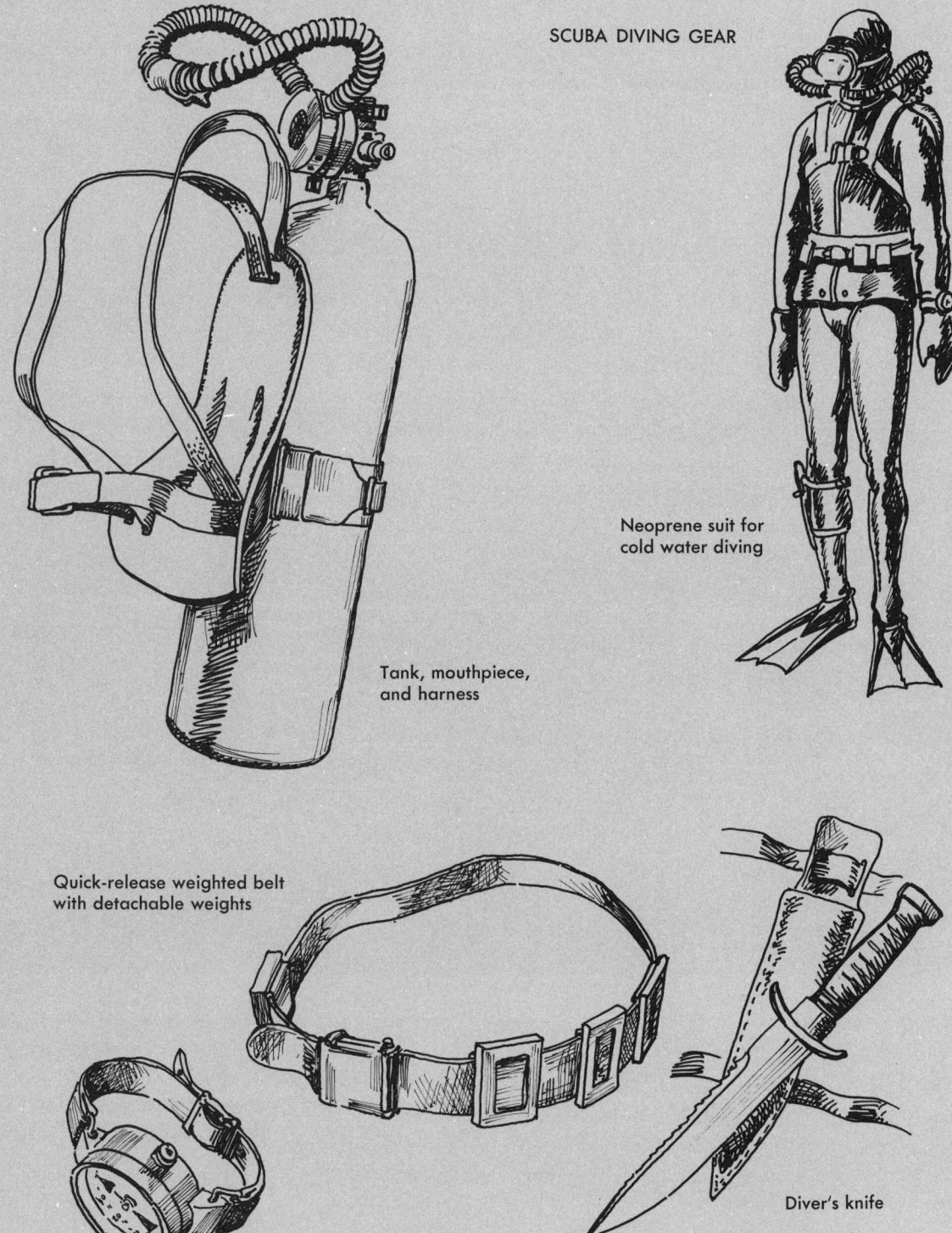

At first, much emphasis was on hunting—with fish spear or spring harpoon. This led to competition for record kills, and general and indiscriminate slaughter underwater soon cleared the available fishing grounds. Inevitably, too, there were accidents, and it was not long before a young diver was killed by a friend's spear gun. Fortunately, the many diving clubs sought to regulate their members' activities, and governments stepped in to set limits on bags and also to set aside certain reef areas as underwater parks.

Now, just as in African safaris, the camera has largely supplanted deadly weapons, while others use their harpoon guns to tag harmlessly certain species of fish whose wanderings and migrations can then be checked. Others hunt sunken treasure or historic wrecks. In many communities, often far from the sea, there are volunteer scuba rescue teams ready at short notice to dive at the scene of an accident.

The free-diving apparatus using compressed air is dangerous below 200 feet, and usually fatal below 300 feet. This is because, as depth and pressure increase, the amount of nitrogen taken in increases also. A diver at 33 feet breathes twice as much air as he does on the surface—otherwise the doubled pressure would cause his lungs to collapse. At 99 feet he breathes four times as much, and at 300 feet ten times as much. The air we breathe is 78 per cent nitrogen, so at great depth the amount of nitrogen breathed is considerable. Nitrogen in this quantity, breathed under pressure, puts the brain centers to sleep. Just how, we do not know, but the fact remains that at great depths, the senses become dulled—a condition known as "nitrogen narcosis," "diver's blackout," or "rapture of the depths"—and the diver eventually becomes unconscious. This danger area can be roughly said to begin anywhere below 150 feet. (At 160 feet, one diver began fumbling to see if he had a cigarette!) Cousteau noticed signs of hallucination just below 200 feet on a test, but safely reached 295 feet. A teammate once reached 305 feet. In a further test the strongest member of the team reached 393 feet, but became unconscious and drowned before he could be hauled to the surface.

This 300-foot limit holds for helmeted divers also. Beyond that, mixtures of gases—usually substituting helium for nitrogen—are used. In 1958, using a helium-oxygen mixture pumped down to a helmeted suit, a Royal Navy diver, George Wookey, went down to 600 feet. A young Swiss teacher, Hannes Keller, "invented" a mixture of gases, and in 1961, with Kenneth MacLeish, then science editor of *Life* Magazine, was lowered on a stage into Lake Maggiore to a depth of 728 feet, at a pressure of some 333 pounds per square inch. In a later experiment—which cost the life of a companion—Keller briefly left and re-entered a diving chamber at a depth of 1,000 feet.

It used to be believed that the human body could not stand pressures at anything like a 1,000-foot depth, and that a free diver, below a certain depth, could not return to the surface against all that pressure. Actually, the human body, except for the lungs, is a fairly incompressible object and, of course, the pressure at any depth is all around, not just down, so that a diver does not feel any "downward" pressure other than that of gravity.

One of the advantages of helium in deep diving is that it is 40 per cent less soluble in water than nitrogen and thus that much less is dissolved in the blood stream. It dissolves quicker than air in the blood and tissues, and escapes faster during decompression. Using helium, the tedious—and sometimes dangerous—decompression stages can be cut by at least one-fourth.

The freedom of action of the scuba diver and his ability to operate at considerable depth—using helium or other gases, such as hydrogen—led undersea-minded people to think seriously about the exploration and eventual exploitation of the shallow portions of the oceans adjacent to the coasts. These Continental Shelf areas extend out from shore to a line where the depth exceeds 200 fathoms, a figure set by international agreement. These areas vary from a few yards in some cases—where the land mass shelves abruptly—to 250 miles out to sea. Their average width is 40 miles,

their average depth (on the flat portions) 35 fathoms, and their total area—world-wide—is estimated at 10,000,000 square miles. This means that vast areas are well within the range of the free diver.

The idea of numbers of men working for prolonged periods underwater led in turn to plans for some sort of dwelling place on the sea floor, maintained at sea-floor pressure and into which divers could come and go—sleep, eat, rest, work, and play—without surfacing, and without need for constant decompression. The first test of one of these was made under Cousteau's direction in the Mediterranean. In September, 1962, two men lived for a week in a steel cylinder 33 feet below the surface. Air was supplied under pressure from the surface. They made several dives daily as far down as 85 feet. In a short test in the same month a diver spent 24 hours at 200 feet in and around a diving bell designed by an American, Edwin Link.

In June of 1963, five men spent a month at 36 feet, while two more spent a week in a smaller "house" at 90 feet. Both crews lived normally, worked, made dives down to 165 feet, and even received diving visitors. A year later, four United States Navy men spent 11 days in Sealab I at 192 feet, despite a constant pressure of nearly 100 pounds per square inch.

Sealab II—a 57-foot long, 12-foot diameter, 200-ton cylinder with laboratory space and accommodations for ten men—was lowered to a depth of 205 feet on August 26, 1965. Aquanauts of the first team stayed down 15 days, diving and performing test salvage work. They were lifted to the surface in a sealed capsule, which was then locked to a decompression chamber. Decompression took over 30 hours. Two other teams were sent down for 15-day stays. No ill effects were suffered, although the first team leader, astronaut-aquanaut Commander Scott Carpenter, was stung by a poisonous scorpion fish. Sea lions, attracted by the numbers of fish surrounding Sealab II, peered into the view ports at the inhabitants.

Another of Cousteau's underwater habitats, Conshelf II, had an underwater hangar into which his diving saucer could be maneuvered and then winched up so that it could be serviced.

There seems no limit to the size or equipment of these undersea homes and laboratories. If man can build 8,000-ton submarines, a sizable complex of semimobile structures of similar size or larger should pose no great problem.

What is the ultimate in pressure which the human body can endure? Men have worked at 700 feet, and reported feeling no different than they did at 200 feet. In pressure chambers—strong steel cylinders into which breathing mixtures are pumped until pressure equals that at any water depth—men have spent many hours at a "depth" of 300 feet without discomfort, although a can of peaches collapsed under the pressure. In miniature chambers white mice have been "sent down" to 3,000 feet, with no ill effects.

In 1969, participating in a project named Tektite, four American aquanauts spent 60 days in, and working from, an underwater chamber at a depth of 42 feet. In 1970, two British scientists made a simulated dive—in a chamber whose air pressure equaled the water pressure found at the corresponding depths—to depths ranging from 1,000 to 1,500 feet. The divers spent five and a half days below 1,000 feet, of which three and a half days were below 1,200 and ten hours at 1,500 feet.

BALLOONS TO THE ABYSS

Until well into the nineteenth century there was a firm belief among almost all scientists that plant and animal life could not exist much lower than 50 fathoms (300 feet). Deep trawls by sailors occasionally brought up strange fish, and naval explorers had fished up worms, sea stars, and other creatures from great depths. But sailors and scientists had little in common in those days, and the latter preferred to cling to their theories.

Then in 1860, the submarine telegraphic cable between Sicily and Sardinia broke. When the ends were fished up from a depth of 6,500 feet, shellfish and polyps were found growing on them—a fact which shook many scientists considerably. In 1869, a scientific expedition brought up living creatures in a deep trawl from over 14,000 feet. Three years later, H.M.S. *Challenger* started out on her famous 3½ year voyage and the science of oceanography was born.

It was natural that, knowing now that life could exist at great depths, scientists should want to penetrate the deep and see for themselves. The first man to do so was an American naturalist, William Beebe. The vehicle he used, called a bathysphere, was designed by engineer Otis Barton. It was a cast-steel sphere, 4 feet 9 inches in diameter, with walls some inch-and-a-half thick. There were three ports of fused quartz, three-inches thick, and two oxygen cylinders provided air for two men for eight hours.

The first descent, in June of 1930, saw Beebe and Barton lowered into the Atlantic, ten miles off Bermuda, at the end of an inch-thick cable. A telephone line kept the two in touch with their barge overhead. The first dive was to 800 feet—a record breaker—and a few days later the two went down to 1,426 feet. Two years later a descent was made to 2,200 feet and, in 1934, to 3,028 feet. The description of the creatures Beebe saw, many of them visible in the pitch blackness by their own strange lights, excited the scientific world.

Later, Barton built a thicker-walled sphere in which he alone descended to 4,462 feet. But 4,400 feet was a mere surface dip

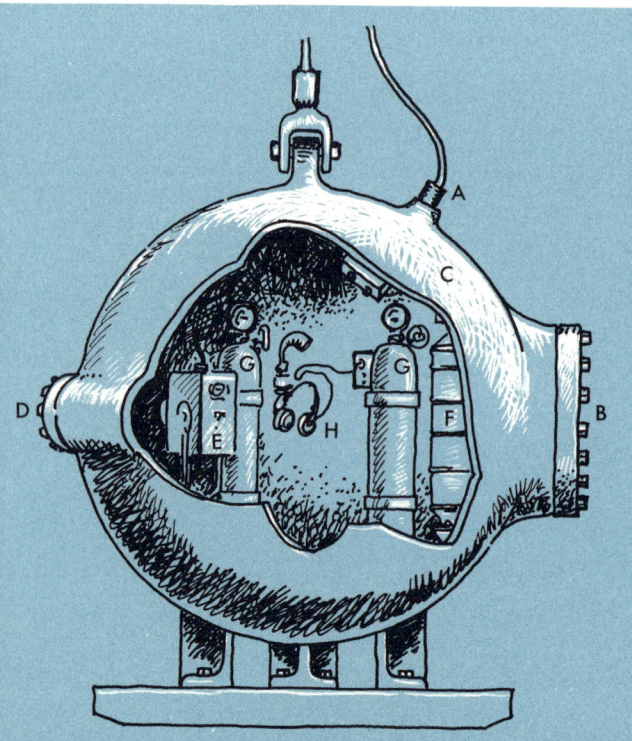

BATHYSPHERE

A — Watertight inlet for electric power and telephone cable
B — Entrance hatch bolted in place
C — 5,000-pound steel sphere, 4' 9" in diameter, 1.5" thick
D — Viewport—there were three, one for searchlight, E
F — Chemical device for absorbing carbon dioxide
G — Oxygen bottles
H — Telephone

compared to the great depths known to exist. Trenches off the Philippines and the Marianas sounded at over 35,000 feet, where the pressure ran to a stupendous 17,000 pounds per square inch! A trawl had brought up a stone with an anemone clinging to it, so there *was* something alive in those huge chasms. The question was, could anything in the nature of a fish live under those conditions? Again, the only way to find out was to go down and see. But the bathysphere, dangling on the end of a steel cable, was not only clumsy and impossible to maneuver, it was dangerous as well. At those depths the pressure is so enormous that the slightest flaw in material would mean death. And to the peril of flooding was added the risk of the cable snapping. It could be tested to hold far more than the weight of the sphere, but a terriffic jerk was imparted when the supporting ship rose on a wave. What was needed was something new—some device which would carry a massive steel sphere gently down and up again, unconnected with the surface, and also capable of traveling slowly over the floor of the abyss.

The man who solved the problem was, of all things, a balloonist. Not so strange, perhaps, because what the inventor, Swiss professor Auguste Piccard, was suggesting was essentially the same device which had carried him higher than man had ever gone. To better study cosmic rays, Piccard had designed a balloon which carried a pressurized steel cabin up into the stratosphere. Just as that container had been buoyed by a substance lighter than air, so another could be made to float down into the depths, held up by something lighter than water.

And that was the origin of the bathyscaphe, named from two Greek words meaning "deep boat." The buoyant substance was gasoline, enclosed in a thin metal envelope or float, and carrying with it not only the thick steel observation chamber but batteries, motors, floodlights, and everything necessary to explore the ocean floor. In the balloon, ballast in the form of sand was released to let the balloon ascend; to descend, gas in the bag was valved off.

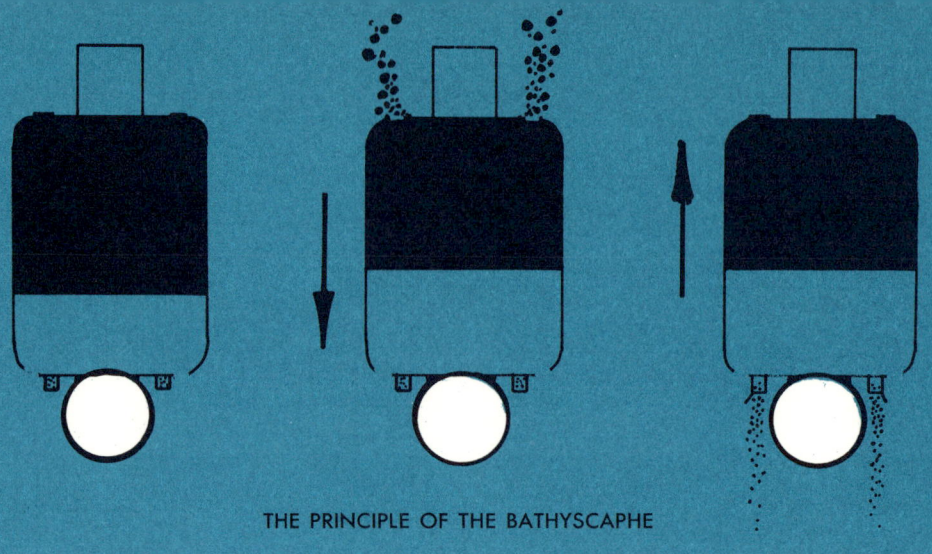

THE PRINCIPLE OF THE BATHYSCAPHE

In the bathyscaphe, ballast of steel shot was released to rise, while to fall, gasoline was valved out. The thin sides of the envelope holding the gasoline were never designed to resist pressure. The bottom of the envelope was open to the sea, and thus the pressure was equalized. The gasoline, being lighter than water, could not escape, and as the water, under terrific pressure, entered at the bottom of the envelope, all it did was to compress the gasoline a little.

By alternately dropping shot—done by releasing an electromagnetic gate—and valving gasoline, the bathyscaphe could be made to rise, fall, or hover. When the sonar instrument reported the bottom near, descent was slowed to a crawl. A chain, hanging down from the bottom of the craft, touched the sea bed, and as the first few links were supported by the ocean floor the bathyscaphe came to rest.

It was rather an odd idea, and the bathyscaphe is an odd-looking vessel. Piccard had had to solve a number of problems. For one thing, his craft had to carry a lot of ballast. As the depth increased foot by foot, so did the pressure; the gasoline was compressed, giving less lifting power—so the bathyscaphe continually went down a little faster. Also, the temperature decreased with depth, and this contracted the gasoline a tiny fraction—so the craft went down a little more. To counteract this, a lot of shot ballast was needed—tons of it. And the ballast release valves had to work properly—because if they stuck, so did the bathyscaphe and its crew. To aid in an emergency, the heavy batteries for the motors which drove the two small propellers and provided lights were carried on the outside of the hull, held by electromagnets. They could be dropped if it was necessary to make a swift ascent. The small sphere, or gondola, of this underwater balloon had a very thick view port and was crowded with instruments—an uncomfortable vehicle for two men to go voyaging in.

The first bathyscaphe was financed by the Belgian Fonds National de la Recherche Scientifique. This government trust had also financed Piccard's stratosphere balloon. Piccard was not romantically inclined when it came to names. His balloon was *F.N.R.S.-1*; the world's first bathyscaphe was christened *F.N.R.S.-2*. It was tested successfully, but many defects came to light and it was decided to build a new float, but retaining the original steel gondola.

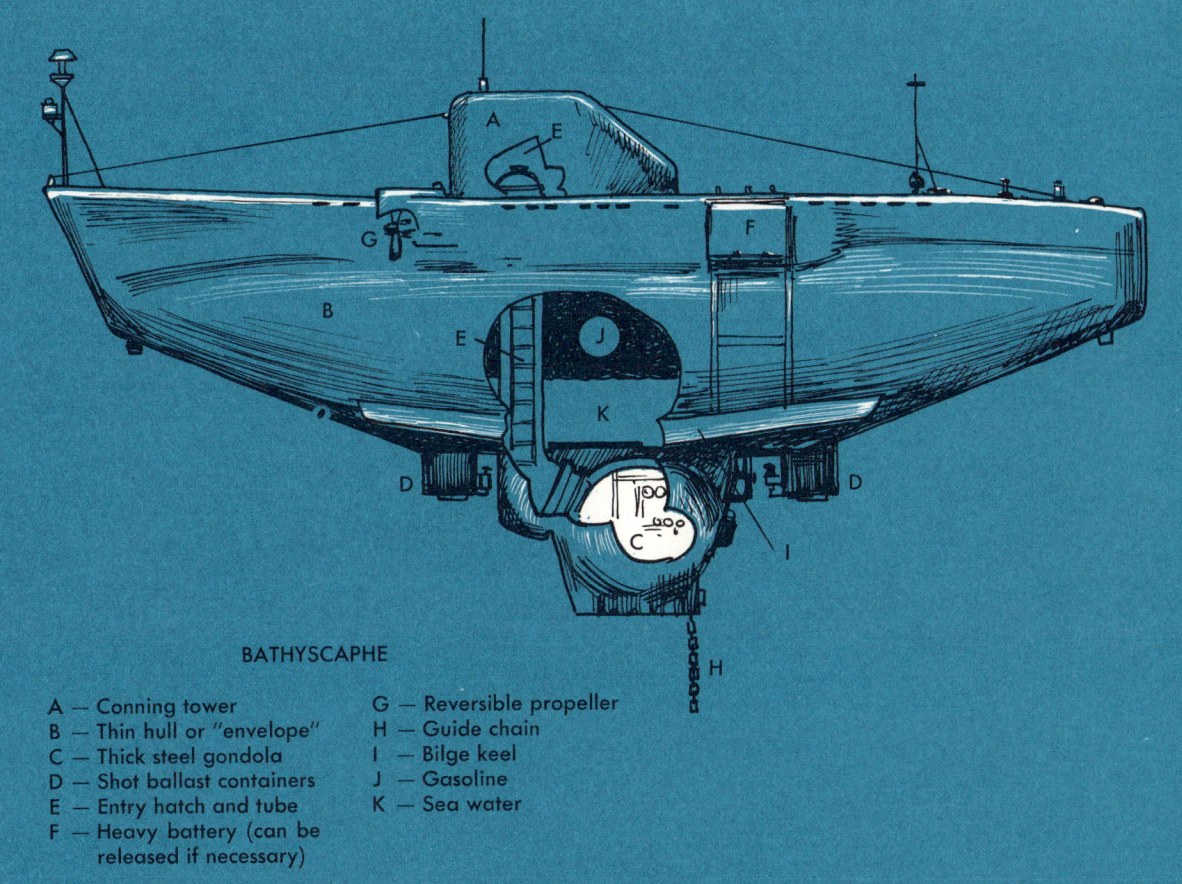

BATHYSCAPHE

A — Conning tower
B — Thin hull or "envelope"
C — Thick steel gondola
D — Shot ballast containers
E — Entry hatch and tube
F — Heavy battery (can be released if necessary)
G — Reversible propeller
H — Guide chain
I — Bilge keel
J — Gasoline
K — Sea water

The French government was now involved, and *F.N.R.S.-3* was built in the French Naval dockyards at Toulon. At the same time Piccard was offered a chance to build another vessel by Italy. This was the *Trieste*—very similar except that her gondola was of forged steel, instead of cast, and thus capable of resisting greater pressure.

F.N.R.S.-3, with Lieutenant Commander Georges Houot and Engineer-Officer Pierre Willm aboard descended to a record-breaking 6,930 feet. This was on August 14, 1953. On September 28, Piccard and his son, Jacques, in *Trieste* touched bottom in a spot 10,332 feet down. In February, 1954, Houot and Willm took *F.N.R.S.-3* down to the ocean floor off Dakar—13,287 feet below the surface. They disturbed a shark on the ocean bed, whose presence there would have astonished the scientists of a hundred years before as much as the arrival of the bathyscaphe did the shark.

Both craft have since made many dives. *Trieste* was ultimately purchased by the United States Navy and on January 23, 1960, with Jacques Piccard and Navy Lieutenant Don Walsh aboard, descended 35,800 feet to the bottom of the deepest known abyss, the Marianas Trench in the Pacific. There, too, a fish swam slowly by—finally answering the question, can a creature of flesh and bone exist beneath seven miles of water?

It was the *Trieste* that finally located the wreckage of the United States nuclear submarine *Thresher*, lying in 8,400 feet of water some 220 miles east of Cape Cod. The photographs taken from the bathyscaphe positively identified pieces of equipment scattered on the bottom as being from the ill-fated vessel.

The superstructure and envelope of *Trieste* have since been rebuilt—using the same sphere—and as *Trieste II*, equipped with mechanical arms and three television cameras, she has played an important part in the Navy's deep-water research program. The French Navy's new *Archimède* has explored the Kuriles Trench off Japan (31,320 feet), and the Puerto Rican Trench, and the Tyrrhenian Sea.

UNDERWATER RESEARCH VESSELS

Not too many years ago, oceanography was a neglected, poor-relation type of science—of little interest to any government, and therefore starved for funds and denied any but the barest minimum of personnel and equipment. But times have changed. The oceans and their riches of fish, minerals, oil—as well as their vital part in national defense—have become so important that comparatively large sums are now spent on oceanographic vessels and equipment. To the many surface ships have been added undersea vessels for peaceful exploration, some of them capable of descending to great depths.

These vessels vary greatly in size and performance—from the 51-foot *Aluminaut*, designed to go as deep as 15,000 feet, to the 9.5-foot Diving Saucer, which does not descend below 1,000 feet. In between are dozens of craft—some designed for exploration, some for underwater recovery, and others for submarine rescue work. Most of the hulls are of steel or aluminum and, almost without exception, are driven by electric motors powered from batteries. Some are light enough to be able to be transported by air.

The uses to which these vessels are put varies with their design and capabilities. *Deep Quest* recovered the flight data recorder from a plane which crashed and sank off Los Angeles. This small object was picked up in 325 feet of water by a manipulator arm on the vehicle.

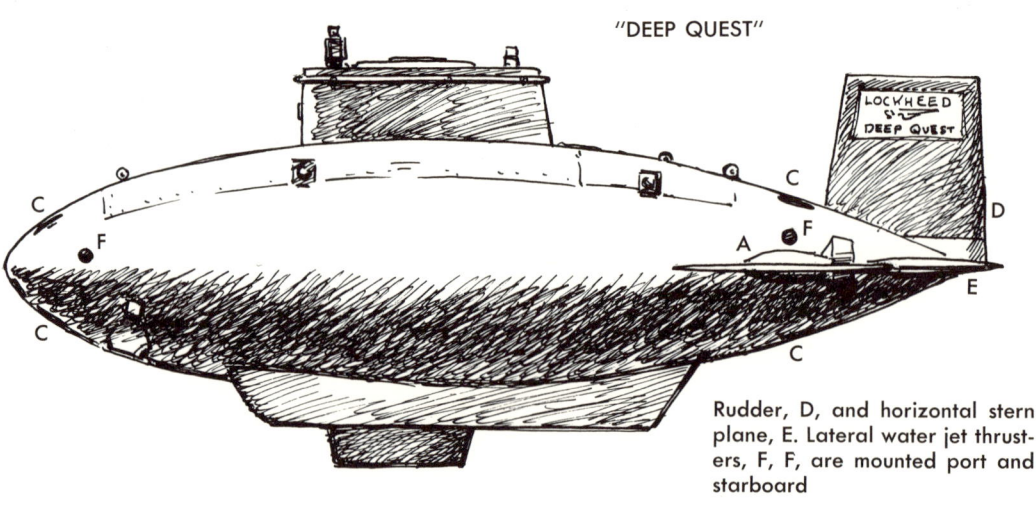

Rudder, D, and horizontal stern plane, E. Lateral water jet thrusters, F, F, are mounted port and starboard

Stern view showing main propellers, A. Top speed: 4½ knots

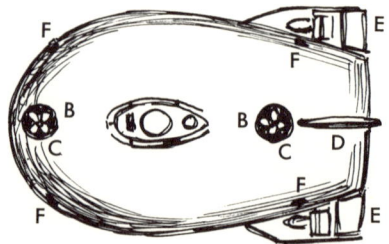

Top view showing vertical thrust propellers, B, mounted in tubes, C, at bow and stern

One of the most spectacular uses of deep submergence vehicles was the finding and recovery of an atom bomb spilled from a United States Air Force bomber which broke apart after a mid-air collision over the Spanish coast in January, 1966. A small fleet of surface vessels assembled. Two deep submergence vehicles—*Aluminaut*, with a 15,000-foot capability, and *Alvin*, which could operate to 6,000 feet—were ordered to the scene as was *Deep Jeep*, capable of working at 2,000 feet. A small submersible, the *Perry Cubmarine PC-3B* for work to 600 feet, was also brought across. More than 150 divers were also on hand, using compressed air to 120 feet, mixed gas to 210 feet, and hard-hat equipment to 350 feet. Sonar bottom-scanning equipment and underwater television cameras were also used.

The methodical searching of the sea bottom in the area in which it was believed the bomb had fallen was a painfully slow process. It was also not without excitement. *Aluminaut* touched bottom at 1,800 feet on the side of a canyon, starting a mud slide which carried her another 800 feet. She surfaced, after dumping her ballast, with 4,000 pounds of mud in her tanks. On March 13, *Alvin*, following a furrow ploughed in the steep slope of the shoreline, sighted the bomb at 2,550 feet.

Aluminaut, able to stay down three times longer than the much smaller *Alvin*, went down within visual distance (with floodlights) of *Alvin* in the pitch darkness. She took down an acoustic "pinger" (a sort of audible beacon), while *Alvin* surfaced to recharge her batteries and mount her mechanical "arm." A line was fastened to the parachute still attached to the bomb, but it fouled an anchor, was cut, and the weapon fell back to the bottom. On April 2, *Alvin* again found the weapon, at 2,800 feet now, and finally with the aid of a complicated device called CURV(Cable Controlled Underwater Recovery Vehicle)—a sort of sled mounting television cameras and controlled from the surface—lines were attached to the parachute and the weapon raised.

Considering that the weapon was wedged on a 70-degree slope

"ALVIN I" (right and below)

Length: 22'   Beam: 8'   Height: 13'
Cruising speed: 2 to 5 knots   Can operate to 6,000 feet

A — Two-man sphere, 7' diameter
B — Trainable stern propeller
C — Rotatable side propellers (port and starboard)
D — Battery space
E — Buoyancy tanks
F — Variable ballast tanks
G — Mercury trim tanks
H — Motors and pumps
I — Scanning radar
J — Conning tower
K — Mechanical arm (not shown in diagram)

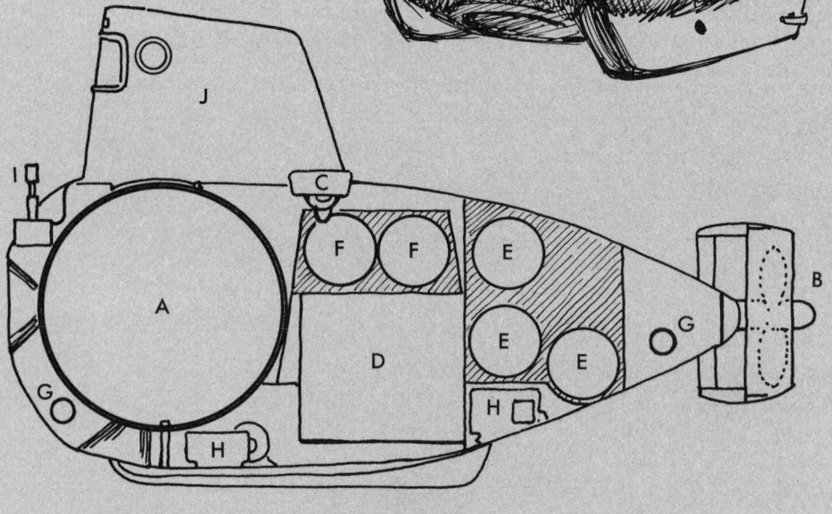

"PERRY CUBMARINE PC-3B"

Length: 22'   Beam: 3.5'   Height: 6'   Crew: 2
Cruising speed: 2 to 5 knots   Can operate to 600 feet

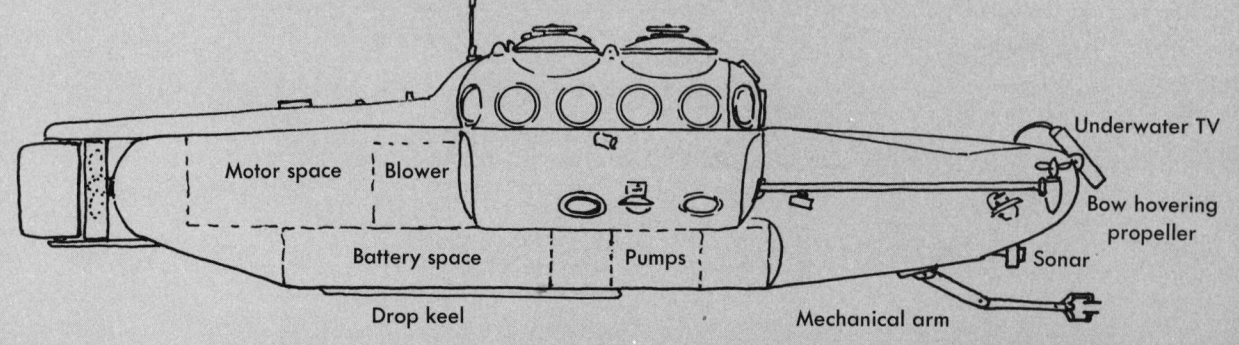

"DEEP STAR 4000"

Length: 18' Beam: 10'
Height: 7' Crew: 3
Cruising speed: 3 knots
To operate to 4,000 feet

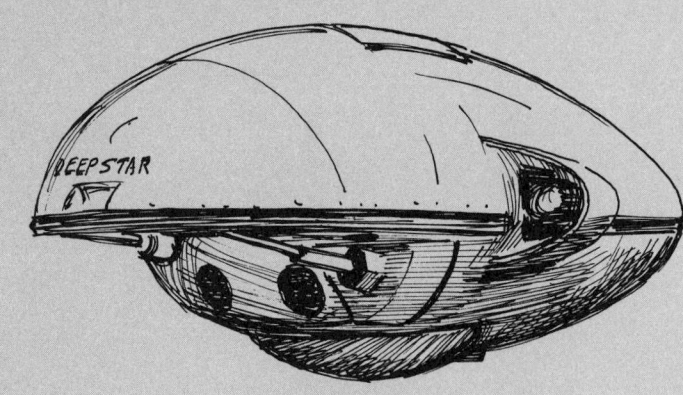

"ALUMINAUT"

Length: 51' Beam: 8'
Height: 14¼' Crew: 4 to 6
Cruising speed: 3 knots
To operate to 15,000 feet

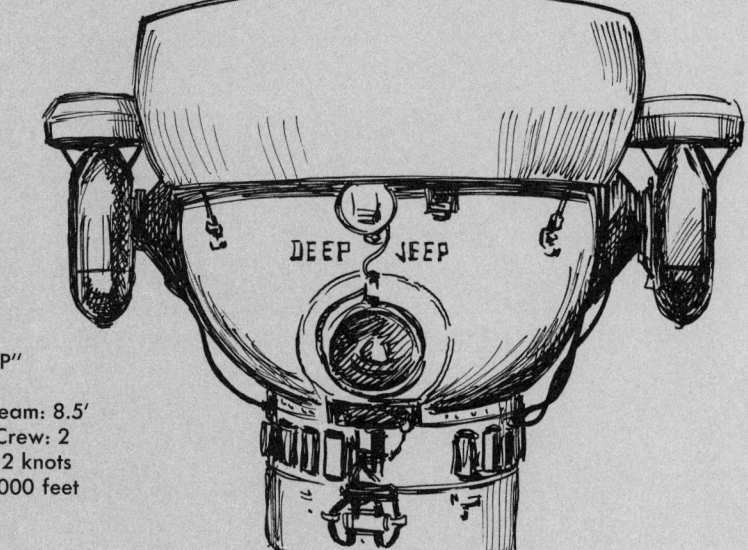

"DEEP JEEP"

Length: 10' Beam: 8.5'
Height: 8' Crew: 2
Cruising speed: 2 knots
To operate to 2,000 feet

at a depth to which sunlight never penetrates, its location and recovery was a remarkable accomplishment.

At a later date, *Aluminaut* rescued the hull of *Alvin,* which sank in a launching accident (fortunately with no crew aboard) in 5,000 feet of water. Once the vessel had been located—and after many tries—*Aluminaut* succeeded in dropping a toggle bar attached to a cable through an open hatch in *Alvin's* hull. When near the surface, divers added more lines for the final haul to the surface.

Deep Diver (able to go down to 1,300 feet) is unique in that she has a pressurized diving chamber from which free divers can operate. When the desired depth is attained, a helium-oxygen mixture of gases is admitted to the diving chamber from the pressure tanks. When the pressure inside equals the water pressure, the hatch in the bottom of the chamber is opened and the divers slip through and out into the depths. When they return, the hatch

"DEEP DIVER"

"DEEP DIVER"

When the boat touches bottom, the diving chamber is slowly pressurized to equal the pressure outside. Then the diving hatch is opened and the divers crawl out. When they are back inside, the hatch is closed and decompression begun.

is closed, and as the submarine vehicle heads for the surface, pressure is gradually reduced and decompression begun. The vehicle is lifted by a special crane aboard her parent vessel and the two-man crew disembarks. The divers, meanwhile, stay in their compartment until decompression is complete.

Off the Bahamas, in 1968, two divers pressurized the chamber of *Deep Diver* to equal a bottom pressure of 700 feet. Then they slipped out of the hatch and for fifteen minutes collected specimens and bottom samples. It is very dark at 700 feet; visibility is between 24 and 50 feet, and the pressure is 326 pounds on every square inch of a diver's body. The work they performed was routine, one of many sorties in a four-week survey. But what was not routine was the discovery of a hitherto unknown current, flowing along the bottom at some ¾ knots in exactly the opposite direction to the normal flow. Not very spectacular, perhaps, but adding another little piece of the puzzle which is the world beneath the surface.

Besides government-built underwater research vessels (URV) and those built by private industry—such as Westinghouse Electric, General Motors, and General Dynamics—there are a growing number of small submarines built for private sport and recreation.

There may come a time when such small submarines are common. There will almost certainly be sight-seeing underwater craft which will take passengers down to tour the spectacular coral reefs and water gardens of our national sea-parks. These small vessels can also contribute to the study of the ecology of the sea. One such, a tiny one-man submarine, was designed by Captain George Kittredge, U.S.N. (Ret.), an ex-submariner. In it Kittredge can dive to 300 feet and has already gathered useful information on the lobster fisheries of his native Maine.

On a larger scale, government underwater research vessels will be able to add to our knowledge of the undersea world on which we must increasingly depend. As our planet's population grows, fish and other products of the sea will become more and more important. If we are not to do great damage by overexploiting these resources, such knowledge is vital.

CHRONOLOGY

333 B.C.	Alexander the Great is believed to have descended in diving bell
375 A.D.	Vegetius, Roman historian, pictures diving helmet of leather with leather hose leading to bladder on surface
1535	Diving bell used to explore wrecks of Roman galleys in Lake Nemi
1620	Cornelius Van Drebbel (Dutch) designs hide-covered wooden submersible
1653	De Son (French) builds clockwork-operated submersible
1690	Edmund Halley's (English) diving bell operated at 60 feet for 90 minutes
1715	John Lethbridge (English) designs cask-type diving suit
1747	Symons (English) builds leather-covered submersible with leather sacks for ballast
1772	Freminet (French) makes watertight leather suit with breathing tube
1773	Day's (English) submersible uses detachable ballast
1776	Bushnell's (American) *Turtle* attacks British ships in Manhattan harbor
1801	Robert Fulton (American) launches his *Nautilus* in France
1819	Augustus Siebe (English) invents bell helmet

1837	Siebe patents complete waterproof canvas suit
1850	Wilhelm Bauer (German) launches *Le Plongeur Marin*
1855	Bauer builds submarine *Le Diable Marin* for the Russians
1860-65	Rouquayrol and Denayrouze (French) invent cylinder for air supply strapped to diver's back
1862-64	Confederates use "Davids" during American Civil War
1863	Charles-Marie Brun (French) launches compressed air-driven submarine *Le Plongeur*
1864	Submarine *El Ictineo* built in Spain by Narcisso Monturiol
1867	Captain Lupuis (Austrian) invents locomotive torpedo
1870	British engineer, Robert Whitehead, sells greatly improved Lupuis torpedo to British Royal Navy
1872	*Intelligent Whale*, designed by Oliver Halstead (American), tested in United States
1872	H.M.S. *Challenger* begins three and one-half year oceanographic voyage
1876	M. Drzewiecki (Russian) builds small pedal-driven submarine
1878	John Holland builds *Holland No. 1* in New York City
1878	The Reverend George Garrett (English) tests small submarine
1879	Garrett constructs steam-driven *Resurgam*
1879	Henry Fleuss (English) invents first fully independent diving suit with mask and tank
1881	John Holland launches his *Fenian Ram*
1885	Nordenfeldt (Swedish) and Garrett test *Nordenfeldt No. 1*, built to use Whitehead torpedo
1885	Claude Goubet (French) constructs *Goubet 1*
1886	Ash and Campbell (English) build first large electrically powered submarine
1888	Gustave Zédé (French) tests *Gymnote*
1888	Spanish naval officer, Isaac Peral, launches submarine powered by electricity
1890	Two Nordenfeldt-type submarines built in Germany
1892	Rudolf Diesel (German) invents heavy-oil engine
1893	Romazzotti launches the *Gustave Zédé* in France

1895	Simon Lake (American) launches his *Argonaut Jr.*
1896	Romazzotti designs the *Morse* with four torpedoes and small periscope
1897	Simon Lake launches *Argonaut*
1897	*Holland VI* is launched
1898	Holland's *Plunger* is launched
1899	Laubeuf (French) launches *Narval* with both electric and steam engines
1900	*Holland VI* is purchased by United States government
1901	Two Russian naval lieutenants, Kolbasieff and Kuteinikoff, build the *Piotr Koschka*
1902	Simon Lake launches *Protector*
1902	A small electrically driven submarine is built in Germany
1903	*Delfin*, first Russian submarine with combination gasoline and electric engines, is built
1906	Germany launches first U-1 submarine
1914	First ship sunk by submarine since American Civil War, H.M.S. *Pathfinder* by German U-21
1915	British liner *Lusitania* sunk by submarine torpedo
1926	Yves Le Prieur (French) introduces improved back-pack for divers
1930	William Beebe and Otis Barton (American) make dives in bathysphere to 800 feet and 1,426 feet
1932	Beebe's bathysphere descends to 2,200 feet
1933	Le Prieur introduces single-glass face mask to back-pack diving
1934	Beebe descends to 3,028 feet in bathysphere
1935	Rubber foot-fin developed by Corlieu (French)
1939	British developed and installed Asdic in surface ships
1943	German battleship *Tirpitz* attacked by British midget submarines
1945	First atomic test
1953	Lieutenant Commander Georges Houot and Pierre Willm (French) descend to 6,930 feet in bathyscaphe *F.N.R.S. 3*
1953	Auguste Piccard and son, Jacques, in *Trieste* descend to 10,332 feet

1954	Houot and Willm in *F.N.R.S. 3* descend to 13,287 feet
1955	Submarine U.S.S. *Nautilus* begins cruise using nuclear power
1958	U.S.S. *Nautilus* makes submerged crossing of Polar icecap
1958	Boatswain George Wookey, Royal Navy, goes down 600 feet in helmeted diving suit
1960	*Trieste*, with Jacques Piccard and Lieutenant Don Walsh, U.S.N., descends 35,800 feet
1961	Diver Hannes Keller (Swiss) descends to 728 feet
1962	First test, under Jacques-Yves Cousteau's direction, in living in a submerged chamber for a week
1963	Five Frenchmen spend month in chamber at 36 feet; two men spend a week at 90 feet
1964	In *Sealab I* four Americans live for 11 days at 192 feet
1965	*Sealab II* is lowered, with space for 10 men at 205 feet below sea level; three teams of men each spend 15 days
1969	In Project Tektite, four American aquanauts spend 60 days at 42 feet
1970	Two British scientists, in simulated dive, spend 5½ days below 1,000 feet and reach 1,500 feet

INDEX

Page numbers in **boldface** indicate illustrations

Air pressure, effects on man, 30, 31, 32
Alexander the Great, 9
Aluminaut, 116, 117, **119**, 120
Alvin, 117, **118**, 120
Aqualung, 102, **103**, 104
Argonaut, 66
Argonaut Jr., **65**, 66
Asdic, 81
Atomic submarines, 93, **94-95**, **96**, 97, 98

Bacon, Roger, 9
Baker's submarine, **61**
Ballast, 14, **17**, 112
Barton, Otis, 110
Bathyscaphe, 111, **112**, **113**, 114
Bathysphere, **110**, 111
Bauer, Wilhelm, 34, 35, 36
"Beaver," German midget submarine, **90**, 91
Beebe, William, 110
Buoyancy, 12, **13**, 14, 47, **56**, 60, 68
Bushnell, David, 18, 19, 20

Caisson disease, 30, 31, 32
Challenger, H.M.S., 109
Chariots, underwater, 77, **91**, 92
Conshelf II, 108
Continental Shelf, 105
Cousteau, Jacques-Yves, 102, 104, 107, 108
CURV, 117

"Davids," Confederate, 35, 36, **37**

Davis escape apparatus, **28**
Day's submersible, 17, 18
Deep Diver, **120**, **121**
Deep Jeep, 117, **119**
Deep Quest, **116**
Deep Star 4000, **119**
Delfin, 55
Depth charges, **80**
De Son's submarine, 10, **11**
Diesel, Rudolf, 71
Diesel engine, principle of, **71**, 72
Displacement, **12**, 13
Divers, early, 7, 8
Divers' equipment (scuba), 27, **28**, 99, 100, **102**, **103**
Diving bells, 8, 9, 20, **21**
Diving suits, 15, 16, 25, 26, 27, **29**
Drzewiecki's submarines, **54**, **55**

Eagle, H.M.S., 20
El Ictineo, **53**

Fenian Ram, **59**, 60
Fleuss, Henry, 27, 28
F.N.R.S.-2, 113, 114
Frogmen, 28, 91
Fulton, Robert, 22, 23, 24

Gagnan, Emile, 102
Garrett, George, 46
Garrett's submarines, **46**, 47
Goubet, Claude, 50, 51
Goubet I, **50**, 51

127

Gymnote, **51**

Halley, Edmund, 9
Hela, 73
Holland, John P. 57-65, 68
Holland No. I, **58**, 59
Holland VI, 62, **63**, 64, 65
Housatonic, U.S.S., 36
Hunley, 36, **37**

Intelligent Whale, 57

Japanese submarines, 86, 87, 88

K Class submarine, **70**
Kaiten I Class suicide submarine, **87**, 88
Keller, Hannes, 105
Kittredge, George, 122

Lake, Simon, 65, 66, 67, 68
Le Diable Marin, 35
Le Plongeur, 37, 38, **39**
Le Plongeur Marin, **34**, 35
Le Prieur-Fernez apparatus, 99, 102
Lethbridge, John, **15**, 16
Link, Edwin, 107
Lusitania, sinking of, 75

M-1 "monitors," **76**
Monturiol, Narcisso, 53
Morse, 52

Narval, 52, 53
Nautilus (Ash and Campbell), 48
Nautilus (Fulton), **22**, 23, 24
Nautilus (nuclear powered), 93, **96**, 97, 98
"Negro," German midget submarine, 90
Nordenfeldt, 41, 47, 48
Nordenfeldt No. I, **47**, 48
Nuclear power, 93-98
Nuclear reactor, **94**

Pathfinder, H.M.S., 73
Peral's submarine, 54
Periscope, diagram of, 68
Perry Cubmarine PC-3B, 117, **118**
Piccard, Auguste, 111, 112, 113, 114

Piotr Koschka, 55
Plunger, **62**, 64
Porpoise, **49**
Protector, **67**, 68

"Q" ships, **74**

Resurgam, **46**, 47
Rouquayrol-Denayrouze apparatus, **26**, 27
Royal George, H.M.S., salvage of, 27

Schnorkel tube, **84**, 85
Scuba diving, 102, **103**, 104, 105
Sealabs, **106**, 107
Siebe, Augustus, 25, 26, 27, 29
SINS, 96
Smeaton, John, 20, 21
Snorkeling, **101**
Sonar, 81
Submarines
   development of, 45
   growth of British service to World War I, 70
   midget type, 77, **86**, **87**, 88, 122
   U.S. R class, 78
   U.S. fleet type, **82-83**, 96
   World War I, 69-78
   World War II, 79-92
Surcouf, 77
Symons' submersible, 16

Torpedoes, 39, **40**, **41**, 42
Trieste, 114
Turtle, 18, **19**, 20

U-boats, **72**, 74, **75**, 82
Underwater research vessels, 115-122

Walter's hydrogen peroxide engine, 84, **85**, 86
Whitehead, Robert, 41, 42, 47

Van Drebbel's submersible, 10

X-craft (British), 77, **88**, 89

Zédé, Gustave, 51, 52